S0-BKL-575

THE GOLDEN AGE OF THE AMERICAN RACING CAR

The Hero

We do not know—we can but deem,
And he is loyalest and best
Who takes the light full on his breast
And follows it throughout the dream.

—AMBROSE BIERCE

All men dream; but not equally. Those who dream by night in the dusty recesses of their minds, wake in the day to find that it is vanity; but the dreamers of the day are dangerous men, for they act their dream with open eyes, to make it possible.

—T. E. LAWRENCE

by Griffith Borgeson

THE GOLDEN AGE OF THE AMERICAN RACING CAR

BONANZA BOOKS • NEW YORK

FOR JASMIN,
My Wife

Library of Congress Catalog Card No. 66-11651

This edition published by Bonanza Books,
a division of Crown Publishers, Inc.,
by arrangement with W. W. Norton & Company, Inc.
C D E F G H

PRINTED IN THE UNITED STATES OF AMERICA

CONTENTS

Part One – The Passion of the Machine

Part Two – Masters of the Machine

Part Three – The Crucible of Speed

Part Four – Fulfillment: The Roaring Twenties

Part Five – Pinnacles of the Golden Age

Part Six – Project Time Machine

Appendixes

ABOUT THE ILLUSTRATIONS:

A large number of the photographs reproduced have never been published before. Where possible, credit has been given to the photographer, but in many cases their names have been lost in time. In such cases the credit line reads GB Collection. Photographs taken by the author are credited simply GB. The initials IMS stand for Indianapolis Motor Speedway.

ACKNOWLEDGMENTS

THE ONLY BOOKS which touch upon the main subject matter of this volume and which have been useful in its preparation are:

500 Miles To Go, by Al Bloemker. Coward-McCann, New York, 1961. This is a masterful treatment of a difficult subject: the history of the Indianapolis Motor Speedway from its beginnings until 1960.

Gentlemen, Start Your Engines, by Wilbur Shaw. Coward-McCann, New York, 1955. A vivid autobiography which is packed with the color of the Golden Age.

Indianapolis Race History, 1909–1941. Floyd Clymer, Los Angeles, 1946. Indispensable to anyone interested in the subject. The 320 large pages are sharp reproductions of contemporary magazine features, mainly from the vast private collections of Jerry Gebby and Joseph G. Wurth.

Auto Racing Winners, Charles L. Betts, Jr., 2105 Stackhouse Drive, Yardley, Pennsylvania 19068. American racing results from 1899 through 1941.

Periodical research for the present volume was done mainly in the Detroit Public Library, with the gracious and enthusiastic assistance of James J. Bradley, head of the Automotive History Collection.

All too little of the history of American automobile racing ever found its way into print, however, and much of that which was not eternally lost survived only in the memories of those who had lived that history or had been otherwise close to it. Hence, personal communication and interviews form the body and soul of this book. Among those who deserve credit for the result are:

Ronald Barker, Surrey, England
Charles R. Betts, Jr., Yardley, Pennsylvania
Al Bloemker, Indianapolis, Indiana
Howard Blood, Sarasota, Florida
John R. Bond, Newport Beach, California
W. F. Bradley, Roquebrune-Cap Martin, A.M., France
Bradford W. Briggs, New York, New York
Russ Catlin, Darlington, South Carolina
Mrs. Suzanne Chevrolet, Detroit, Michigan
John Christy, Hollywood, California
Erret Lobban Cord, Reno, Nevada
S. C. H. Davis, Guildford, England
Pete de Paolo, San Francisco, California
Dick Doyle, Palmdale, California
Mrs. Edith Savage Duray, Twentynine Palms, California
Leon Duray, Twentynine Palms, California
Reeves Dutton, Los Angeles, California
Robert Fabris, Milpitas, California
Earl Fancher, Alameda, California
Scott Fenn, Inglewood, California
Jerry Gebby, Dayton, Ohio
Leo W. Goossen, Los Angeles, California
Ben F. Gregory, Kansas City, Missouri
Ray Harroun, Indianapolis, Indiana
Mrs. Gwenda Hawkes, London, England
Roscoe C. Hoffman, Detroit, Michigan
Jerry Houck, Yucaipa, California
Bill Kenz, Denver, Colorado
Gerald Kirkhoff, Glendale, California
Karl Kizer, Indianapolis, Indiana
Frank Kurtis, Glendale, California
Mrs. Carrie Lockhart, Glendale, California
Dick Loynes, Long Beach, California
Karl Ludvigsen, Pelham Manor, New York
Charles Lytle, Sharon, Pennsylvania
Jean Marcenac, Burbank, California
Pierre Marco, Molsheim, France
T. A. S. O. Mathieson, Cascais, Portugal
Eddie Meyer, Los Angeles, California
Louis Meyer, Los Angeles, California
Eddie Miller, Sr., Culver City, California
Mrs. Edna Miller, Los Angeles, California
Tommy Milton, Detroit, Michigan
Fred E. Moscovics, New York, New York
Fred Offenhauser, Los Angeles, California
Eddie Offutt, Los Angeles, California
Ernie Olson, Sherman Oaks, California
Art Pillsbury, Beverly Hills, California
Laurence Pomeroy, London, England
Proto Tool Company, Los Angeles, California
Roy Richter, Bell, California
Walter F. Robinson, Jr., Bellevue, Washington
Walter Sobraske, Los Angeles, California
Gordon Schroeder, North Hollywood, California
W. Frank Sharp, Jacksboro, Tennessee
Clay Smith, Long Beach, California
Ralph Snoddy, Hawthorne, California
Wayne Thoms, North Hollywood, California
Bill Tuthill, Daytona, Florida
Cornelius W. van Ranst, Dearborn, Michigan
Zenas Weisel, Los Angeles, California
Bill White, Inglewood, California
Ted Wilson, San Jose, California
Ed Winfield, San Francisco, California
Joseph G. Wurth, Detroit, Michigan
William B. Ziff, Jr., New York, New York

PART ONE

The Passion of the Machine

(Left to right) Louis Chevrolet, Harry Miller, Fred and Augie Duesenberg. A most remarkable photo, made in 1923. IMS

chapter 1

THE PASSION OF THE MACHINE

RACING OBVIOUSLY is madness. For one thing it is quite lethally dangerous. For another, hardly anyone makes a penny out of it, and most of its participants live lives of real sacrifice in order to be part of it. But this is nothing special in itself. There is no end of strange pursuits that give large numbers of men—and often women—challenges against which they can test their mettle in an age from which high adventure has largely vanished. And there are hordes of people—artists, musicians, poets, writers, ski bums, and sun and surf bums among them—who pursue their passions as though economic reality did not exist. Then too there is the vast, square, solid citizenry and its dronelike existence, and there are other aspects of the "respectable," the "responsible," world that seem equally mad from other points of view. Perhaps we just choose or get stuck with our personal madnesses. In motor racing, however, everyone chooses his fate, and when a driver crashes and burns, the standard philosophic reaction is, "Terrible. A rotten, lousy break. But nobody held a gun to his head and made him do it."

Why *do* they do it, then? There is much more to it than the "because it's there" that silences dull souls who are mystified by mountain climbing.

One element, certainly not exclusive to motor racing, is the gamble for the ultimate in stakes. "Dicing with death" is the British cliché for it, and death wins the toss all too often. Along with this gamble is the gamble for victory against all one's competitors and the heady atmosphere of anxiety and aspiration which it creates. The delirium of victory in this esoteric form of gladiatorial combat is utterly beyond the hopes of most of the contestants, yet they remain in avid competition. It is a strange drug indeed.

It is compounded of the emotional thrill of the whole experience—from the conception of a design, to the drawing board, to the patterns and forgings and machine work and assembly. Then to the dynamometer and then, finally, to the track. The clout on the back of tremendous acceleration; the exhilaration of blazing speed, hurtling through the wind crouched in an open cockpit; the thrill of the sound and smell and of the machine's savage and, the driver hopes, exquisite response to his slightest wish or command; and much much more. There are plenty of men who will do almost anything to get a ride in a race car, with money the farthest thing from their thoughts.

A very real part of the thrill is sexual, a statement which will not come as news to most psychologists nor to most members of the racing fraternity. There are those of us who very consciously can experience degrees of orgastic transport with a suitably inspiring mechanical companion.

People with different sexual temperaments are attracted by different types of machinery. The oval-track race car (there's a phallic nickname for it in the "in" jargon) attracts a remarkably uniform type. The flat-out (he says "balls out") stand-on-it-and-turn-left charger is typically a person of very strong, virile libido. It seems to be and probably is an occupational prerequisite.

Then there is the attraction of the purebred racing car as the incarnation of perfection of design, workmanship, efficiency, function. The denizens of that world—drivers, mechanics, car builders, car owners, and devout followers—all sense this, each in his own way.

Many of them see the machine also as an art form, as sculpture in metal to whose three static dimensions are added those of movement, sound, and almost-organic life. Language is full of anthropomorphic expressions such as, "its engine burst into life." For the real consort of the machine such expressions are more or less literal, not merely picturesque. It lives, throbs, is faithful, then fickle, has emotions, and elicits them. It can seem to be more than sentient. And it has its life cycle. When we put it to rest we often mourn for a dear companion who has gone on.

The machine is an artifact of our busy civilization that is no less a vehicle for soaring flights of imagination than those that are conventionally recognized as such: painting or architecture, or, less conventionally recognized, such really lyrical disciplines as pure mathematics and celestial mechanics. There is some measure of beauty, logic, and harmony in any machine that works. But those machines which are the embodiment of man's most inspired thinking in the idiom of technology stand as monuments—I will not say to man's conquest of nature, a dangerous illusion—but to the discovery and realization of the potentialities that nature conferred upon him before turning him loose upon his mote in the Universe. Art, science, technology, philosophy, and poetry are only artificial categories for coping more conveniently with the astonishing creative processes of man. All are parts of the same endowment.

Many people who knew Harry Miller state that he was much more artist than engineer, and he probably would have been the first to admit the fact. The brochures of Harry A. Miller Incorporated boasted of the firm's *Creative and Research Division,* under the personal direction of the chief. Edna Miller stated that whenever her husband completed a new racing car he brought her to the plant to see his latest creative, esthetic triumph. The love between them was both tender and strong, and Harry's cars were, among other intense things, songs of love.

Naturally, everyone with a hobby or passion would like to find ways to enable it to support him financially. Chevrolet, Duesenberg, and Miller all did their utmost to make racing serve as a bridge to the manufacture of passenger cars and aircraft engines, where the real money was. The big money never came their way, and what money did come was quickly lost because they all lacked the required practical and calculating mentality. Instead they all made money for much shrewder men. Louie, Fred and Augie, Harry, and most of the entire cast of wildly, joyously inspired and fearlessly adventurous men who created and acted out American racing's golden age crossed the final finish line with little more than memories for wealth. But what memories they were and forever will be!

chapter 2

THE CRADLE OF SPEED

PERHAPS THE GREATEST single leap in the industrial and technological development of the United States was triggered by the First World War. Of course the substructure on which the whole edifice was raised had been evolving at a good rate for decades. Still, aside from a few urban centers, it remained a wild, woolly, and provincial country. From the pinnacle of present-day development it is almost impossible to visualize how primitive conditions were in America such a short time ago. The backwardness was general, but our concern here is specific: the automobile and the development of big-time thoroughbred racing.

Except for urban centers the United States was devoid of roads that motor vehicles could use in any reasonable, practical sense. True, there were wagon tracks that animal-drawn vehicles could negotiate at a plodding pace, and here and there a good horse-drawn rig with buoyant suspension could stir up the dust. Decent roads ceased to exist about fifty miles west of New York although it was possible, if terribly arduous, to drive to Chicago in the warm months of the year. In winter the trip was outright impossible. And west of Chicago road conditions got worse. Whenever or wherever rain fell or snow melted there was the terror of being trapped in seas of mud. Light, buggy-based vehicles such as the Orient Buckboard and the Curved-Dash Olds were the early equivalent of the go-anywhere Jeep, followed by Henry's Model T, and racing categories were built around them. Heavier, more powerful, faster cars were confined almost exclusively to urban areas. They raced, too, chiefly on horse tracks that were like plowed fields and, occasionally, on one of the few road courses that the still very young country had to offer.

This helped to check the development of the American automobile and to channel it narrowly. It discouraged the development of the sort of high-performance machines which had already become common overseas. True, there were occasional one-off cars built specifically to break straightaway records and even, rarely, to participate in the rare, big road races. But what in those days was called The Movement—a term which summed up the movement for motorization, for reasonable roads, for mechanical and, therefore, industrial and economic progress—was far behind that of France, Germany, Britain, and even Italy. It had to fight the apathy and resistance of tradition, the open hostility of the farming element, and the militant opposition of the railroads. The railroads fought tooth and claw against motor roads. "We have brought the nation in a flash," they argued, "from the covered wagon to efficient modern transport. These wealthy owners of motor cars who, in their own selfish interest, are agitating for the construction of motor roads with public funds could ruin the country if they were taken seriously." Still, such desperate but well-planned propaganda efforts as the Glidden Tours communicated The Movement's message to millions of Americans. But it was not until after World War I and its accompanying lessons of the military importance of mobility that any significant progress began to be made in the creation of such a road system. Obviously, it was one of the most basic factors that influenced the phenomenal destiny of the automobile in the United States.

It was during that same crisis that the aeronautical arts and industries received massive patronage from the government for the first time. The aircraft engine was derived directly from the automotive engine, and both fields flourished, thanks to military needs, contracts, financing, and urgency. It was a major turning point in technological history.

For example, American metallurgical practice was so far behind that of Europe that, before the war, most of the better makes of American automobiles depended upon European suppliers for their critical metal components. Packard, Peerless, and Pierce-Arrow had their cylinder blocks cast in France and their crankshafts forged there. Simplex proudly advertised its use of Krupp steel from Germany. The automobile was still too new in America and too unimportant a market to persuade the giant steel companies that they should invest in the production of alloys other than the old standbys that were good enough for all the rest of American industry. Similarly, none of the big foundries were interested in changing their traditional, "good enough" casting procedures just to satisfy the alleged requirements of this upstart industry whose future was still unclear. If axle and drive shafts twisted like corkscrews or snapped, if castings with irregular wall-thicknesses failed, it was the fault of the automotive designer. The remedy was simple enough: increase the safety factor; beef it up. This is why Miller shipped his patterns five hundred miles away and Duesenberg shipped his twice that distance to small, artisan-minded foundries. They were slow and costly, but their handmade products were precise, uniform, and dependable.

Fine steel alloys were available too but, again, only from very small, specialist suppliers. Fred Offenhauser recalled that their uniform quality was far superior to that of the same alloys when they began to become available from the major steel mills. These are some of the problems that the builders of the pioneer thoroughbreds had to live with and overcome. They walked onto a darkened stage. When their work was done it was incandescent.

In the beginning the automobile was a marvelous novelty that the pedestrian and equestrian hordes would pay well to see perform, as every horsetrack and fairgrounds operator was quick to recognize. The spectacles of speed began with individual barnstormers and stunt artists whose solitary performances were as thrilling to city folks as they were to yokels. There were hosts of these showmen, and among them were Carl Fisher, who later founded the Indianapolis Motor Speedway, and the granddaddy of all kings of speed, Barney Oldfield.

After the barnstorming artists came the organizers of motorized circuses: whole stables of cars that went wherever the money was and ran rigged races to thrill and mulct the naïve public on a wholesale basis. Alex Sloan's troupe was the most famous of these, and it gave the United States and the world many of the best drivers of all time. Tommy Milton recalled:

> It was 1913. I was just a kid who wanted to race and I happened to have my own Mercer. Mercer had a very good name just then and once when Sloan brought his show to St. Paul he offered me a job with his fakeroo outfit. All I had to do was furnish the car, maintain it and drive it and he would pay me $50 a week. I took the offer and stuck it out with Sloan for three years.
>
> He had a cattle car with decking in it, which made it possible to haul nine race cars wherever there was a railroad. As soon as we left my home town and hit the open road he cut my salary to $35 a week; that was his style. There was one eight-day period when we travelled 3000 miles and raced in five cities. For $35 a week. No prize money.
>
> My Mercer was just a stock model. I never had a chance of doing anything against the real race cars in the outfit, so I never was given instructions. Louis Disbrow was the big star during most of that period and it was usually arranged that he would win. All the dirt-track promoters ran these phony shows.
>
> One day we raced in Peoria, where I learned never to boo anybody. My sister and her husband lived there and they came to see their daredevil kid-kin in action. I didn't stand a chance anyway but those bums in the stands roaring at me, "Get off the track, you lousy coward!"—and a lot worse—got to me.
>
> It sickened me with being slow. I chopped my car's frame and shortened its wheelbase from 108 to 96 inches. I installed a 450 cubic inch Wisconsin T-head engine—which is what Stutz used in those days—and I suddenly had a darned good track car. We wound up the 1915 season at Shreveport, Louisiana. I was tired of getting beaten, so I beat everybody, including Disbrow. Sloan fired me,

putting an end to one of the dark pages in my history, and in 1916 I joined the Duesenberg team.

This is one of the ways that racing drivers were born in that rugged age. They had their own cars, and it was their cars as much as any personal talent that gave them their entrée into the game. Others had to take a harder route. Duesenberg driver Eddie Miller's case was typical:

> I was twenty years old and had landed a job in the test house of the Elcar Company in Elkhart, Indiana. It was in the fall of 1915 that I wrote to the Duesenberg brothers, in St. Paul, saying that I'd like to work for them. Fred replied saying that they didn't have any money but that if I happened to be at Elgin, Illinois, for the road race there, he'd be glad to talk with me.
>
> I showed up at Elgin and they let me help watching the wheels on the back-stretch. In those days they used to have wheels spotted around the course against the inevitable blowouts. That was my start in racing—watching a wheel. So I went to St. Paul, worked for Fred and Augie and became a mécanicien. That's how you had to start unless you had a lot of money to buy a car.
>
> In those days the cars were all four-bangers with up to 500 cubic inches displacement. We had no more than five-to-one compression ratios but with those huge bores and strokes it took a real man to crank one over.
>
> Pretty good cars, like the Stutzes, would get up to about 105 MPH and then start to shake and that was the limit of how fast you could go. Still, they were pretty reliable. The Duesies were a little bit faster but they really were lacking in some areas, like the brakes. Their brakes were just no damn good. It took a lot of downshifting to second gear to get slowed down and there wasn't anything like a close-ratio gearbox; the spread was about thirty per cent!
>
> We used to think nothing of racing for six and a half hours. The pounding was so bad that you had to wear a corset. At the end of a race you'd be black and blue, unable to breathe and you would have lost eight or nine pounds. To win a race, then, with the wheels and the hubs and their weight and the tire size and the steering and the weight of those clunkers and the way the springs worked, it took a good man even to ride in the things, just to get hauled around. Experience really meant a great deal and it was rough to be a rookie. Loads of us stacked up in our first races. Lots got killed and lots of others never got beyond the hard-luck stage, just couldn't break the barrier. You either did it or you didn't.
>
> And can you imagine the life, getting around the country by train? Five days to get from Newark to Los Angeles in wooden coaches going 30 to 40 MPH, crawling 3400 miles across the States. Getting *anywhere* was a terrific job.
>
> And the technology, or the lack of it. Who knew anything about spark-plug ranges in those days? Nobody. About gasoline? Nobody. You knew that the engine would knock, but that was all you did know. You knew it was making

a noise but you had no idea of how to get rid of it. Solving it took a long time.

Each oil company blended its primitive fuel differently—anything that would start and run and not knock too badly. Nobody had any conception of octanes and we used to test gas with a hydrometer, hoping to detect something about its quality. Some producers back in the Teens sent us a couple of barrels of alcohol and asked us to try it. Everyone said it wouldn't work and we used it to wash parts in. It was just too drástic a change for that far back. It's easy to be very conservative when you're running on your last few bucks.

Then there were the tires. When the 300-inch cars began running and going pretty fast we were still running on clinchers. Can you imagine that? We had thirteen Schrader lugs, plus the valve; fourteen holes in the rim just to be sure the tire was going to stay on it—you hoped.

They were eight-ply tires and we ran between 90 and 100 pounds of pressure in them. I remember a tire change when a mechanic said, "Well, that does it!" and hit the tire with his hammer. The hammer flew up, cracked him on the head and laid him out cold. That's how much rebound there was in those things.

Then came the straight-side tire—I think Goodyear gets the credit for it. When I began driving I would have to take a walk when the mechanics were mounting my tires. I couldn't stand the sound of the beads breaking as they pried them over the rims with big crowbars.

Look at the problems of space travel today. It was like that for us. You couldn't go down to the parts store and buy a rubber cup to put in a brake cylinder because there weren't any. The Duesenberg hydraulic brake was superb for its time, but it wasn't arrived at easily. Imagine! You didn't have hydraulic brake fluid. You knew nothing about the mass-per-cylinder relationship. We had to make our own brake cups out of leather. We had to find a fluid that wouldn't freeze, rust, or get hard or sticky. But except for our not having servos on both brake shoes, those brakes were as modern as the industry's best forty years later.

The pioneering days were difficult but there were certain advantages to the economic atmosphere. Sponsors were abundant and quite open-handed in their efforts to get their products proved in the ultimate test of big-time racing. Makers of tires, spark plugs, fuels, lubricants, components, and accessories supported the sport in generous style. Purses were fat and quite abundant, and the economic organization of the sport helped it to be strongly self-supporting. Art Pillsbury said:

The boys ran a lot of races, for a lot of money, but not like they do today. After the first war there were always at least ten first-class tracks in addition to Indianapolis. Every one of them had two races per season, many of them three. The schedule set up by the AAA Contest Board based the purse for any major event at $100 per mile. There usually were from twenty to twenty-five 250-mile races in the course of a season so, with purses of $25,000 each, there was quite a bit of money to be made and of course the crowds we drew were tremendous.

I could be hung for making this statement but I know I'm telling the truth.

There is nobody today who could build the sort of speedways we did and survive the present setup. There is no group of stockholders that would give forty per cent of their gate to any group of racing drivers. They couldn't do it; it would never pay out. The only reason they're getting by at all now is that they are using tracks that were built by others in the past. The only reason Indianapolis is what it is is that Tony Hulman doesn't want any of that money.

The difference between the old days and these days is that now they operate on the theory that one poor race car can support a mechanic and his family and a driver and his family and also give the car owner some profit. It never has been possible and it never can be, even *with* the forty per cent slice. In the old days the owner didn't have to have that money to support the car. The driver didn't have to have it to eat on and the mechanic didn't get paid so much per day or week. Those men got paid when their cars won and if you didn't place in the money you didn't make any money.

It was professionalism all right, but it was very different from the kind which followed. Right through to the end of the Twenties American racing had aristocratic associations which it has not had since. The fraternity itself was full of men of substantial wealth and culture. Cliff Durant was just one example. There were so many others. Joe Boyer, whose father owned the Burroughs Adding Machine Company. Eddie Hearne, an heir to a great gold-mining fortune.

And then there were all the men like Milton and de Palma, the most polished of gentlemen and brilliant adornments to any gathering. So it was the accepted thing that the best families hosted these people, because they were the same type of people. Mansions, filled with all the luxuries that wealthy people surrounded themselves with in those days before taxes, were the natural habitat of the racing fraternity. When you invited one or a hundred of these men to a dinner, a banquet or a party you knew that each of them would appear in evening clothes and conduct himself well. Of course the grease-monkey element was part of the scene but the tone of the sport was set very largely by people of means and refinement.

The AAA Contest Board was composed of men whose feeling for the sport, and their private means, made it possible for them to pay well for the privilege of their positions. What it cost Eddie Rickenbacker or Art Herrington to serve as chairmen of the Board, just in annual travel expenses, was a lot. The last thing any of us wanted out of the Board was monetary gain. It probably cost me two to three thousand dollars a year just to be a member and no chairman ever spent less than five to ten thousand a year in doing his job. We got into the sport because we thought it *was* a sport.

It was about 1920 that we got the Board properly organized, with Rick in command. Everyone was treated the same. There was no favoritism. None of us could accept the simplest gift. We purged the fraudulent element and no matter how much our decisions might be disliked no one ever could charge us with an ulterior motive. Maybe we were wrong, because things have changed and they seem to be working. But we believed that the sport should be governed by people who have no vested interest in it—if it is to be a sport.

W. F. Bradley, a young English journalist, settled in Paris in time to cover the catastrophic Paris-Madrid Road Race for *The Automobile* of New York. The name eventually changed to *Automotive Industries,* but Bradley was the European correspondent for these magazines continuously until 1956. He is one of the finest of all automotive writers, and he participated in much of the important history of the automobile.

In 1913 C. A. Sedwick of the young Indianapolis Motor Speedway sought Bradley out in Paris and appointed him European representative for the track. Then, when the American Automobile Association (AAA) Contest Board was overhauled and given its real strength just after World War I, Bradley was made its delegate to the International Sporting Commission of the Association Internationale des Automobile Clubs Reconnus (A.I.A.C.R.) in Paris, now called the Féderation Internationale de l'Automobile (F.I.A.). Thus, Bradley was in the middle of most intercontinental racing activities for about a quarter of a century, including their real beginning:

> Sedwick wanted European talent in the Indianapolis race. The cars he wanted most were the Peugeots. We went around and saw them. They were entirely ignorant of conditions in America. George Boillot was the spokesman for the team and he thought we were nothing but a bunch of Buffalo Bill types racing on dirt tracks where the organizers were in the habit of running away with the prize money. There were plenty of these true stories going around.
>
> The fact that there was a fine track at Indianapolis, run by wealthy men who were not trying to make money from it but who wanted to help the industry and the racing movement was totally unknown in Europe. We had the greatest difficulty convincing them that the impressive prize money was already in the bank and that it definitely would be paid. Finally Boillot agreed to let Goux and Zuccarelli go over, taking a couple of the previous year's cars.
>
> These Europeans knew nothing of oval-track racing nor of driving in traffic, which they did not like. De Palma watched their troubles on the first day of practice, came to their pit and said, "There are tricks to driving here that you boys don't know. You just tuck in behind me and I'll show you the groove." He did just that, being that kind of sportsman.
>
> The good conditions, the impartiality, the helpful spirit, all impressed the Frenchmen. During practice they had to put up with a lot of nerfing by a few American drivers who wanted to scare them off the track. AAA observers on the circuit spotted this and told the offenders to quit it or get off the track themselves.
>
> Came the race and Jules Goux won it. He was paid his staggering $20,000 the same day, plus other prizes. The checks had been made out in advance so that it was only necessary to fill in the names. Goux returned to France nine-tenths

American—his booty including the hand of the prettiest girl in Indianapolis—and American racing was *made* in European eyes. The "500" remained the Mecca for the Europeans until they finally became hopelessly outclassed. And, of course, the specialized American cars became equally unsuited to European road racing.

It was a world of a very different time sector, and it is hardly comprehensible to us now. Men wrote heroic verse about the "knights of the roaring road" and these were regarded as modern heirs of the gladiatorial tradition. In the popular mind, their daring was a match for that of the bravest ace who pumped lead between the propeller blades of a Spad biplane to the song of screaming guy-wires. It was a most historic and significant adventure, a penetration of uncharted seas, without instruments, which helped to bring us to where we are today. It was a unique epoch, marked by dazzling human achievement, and it ended with the 91.5-cubic-inch formula and with the year 1929. That particular human epoch ended, but there was not a ripple in the evolution of the machine. No substitute ever was found for quality, for class.

chapter 3

THE COURSES

THE GOLDEN ERA of thoroughbred automobile racing began almost imperceptibly in the early Teens of the century and ended abruptly with the Twenties. It and the Speedway Era, which put much of the roar into The Roaring Twenties, were one and the same.

The story of the Indianapolis Motor Speedway has been told definitively and skillfully by Al Bloemker in his book *500 Miles to Go*. Racing on public roads was being systematically outlawed in nearly every state of the Union, and the automotive industry was without any proper high-speed proving ground. Ex-barnstormer Carl Fisher had become prosperous in the industry (Prest-O-Lite acetylene lighting equipment), still loved racing, and conceived the idea for the original speedway.

Fisher was not interested in direct profits but in helping The Movement, which included the industry of which he was a part. The industry at that time and until local shortages of coal, iron, and investment capital drove most of it north was centered to an important extent in and around Indianapolis. Fisher found ready encouragement and backing for his idea. He launched it in 1908, the first race was held on the 2.5-mile oval in 1909, and the first "500" took place in 1911. The Speedway is still the most successful motor racing plant in the world, thanks to consistently enlightened management and consistent dedication to the sport above all monetary or other ulterior considerations.

The speedway concept was a simple and logical adaptation of the typical high-

As originally conceived, the Indianapolis track was to have been a combined speedway and road-racing circuit. IMS-GEBBY

quality horse-racing plant to the needs of the automobile. Others were quick to pick up the idea. In the spring of 1909 Asa G. Chandler and Ed Durant began the construction of a two-mile oval at Atlanta, Georgia, which they surfaced with local gravel. Simultaneously out on the West Coast Fred E. Moscovics hatched an idea which was to rock the world.

He had received his technical education on both sides of the Atlantic. One of his first jobs was in the drafting office of Wilhelm Maybach, with Daimler Motoren Gesselschaft. Then he returned to the United States, where he went to work for the New York importer of Continental tires from Germany. Through this connection he was called upon to manage the Daimler-Mercedes team which competed in the first Vanderbilt Cup Race in 1904. He then settled down to full-time engineering work and in 1907 was engaged by the newly formed Allen-Kingston Motor Car Company of New York to design a new, powerful car to bear that name. Among the little firm's employees were a dashingly handsome young test driver named Ralph de Palma, and it was on one of Moscovics' machines and under his tutelage that de Palma got his first racing ride in the Briarcliff Trophy road race in Westchester County, N. Y., on April 24, 1908.

In 1910 Moskovics became general manager of the Remy Electric Company, in 1912 commercial manager of Nordyke & Marmon, and in 1923 president of Stutz. He was one of the earliest members of the Society of Automotive Engineers, played a truly outstanding role in automotive history and in racing, and did outstanding engineering work for the United States Air Force during World War II. When I last saw him, in 1955, he was seventy-six and dynamic. He hurtled daily between his residence in Connecticut and his engineering office in New York in his Mercedes-Benz 300 SL gull-wing coupé.

When Moscovics was a young student at the Armour Institute in Chicago and at the Polytechnic in Zurich (spending most of his holidays in Paris, where the cars were) his favorite hobby was bicycle racing. And in the States and in Europe he came to know most of the celebrities of that sport, including the one-time world's champion, Briton Jack Prince.

When Prince's athletic years were behind him he came to the United States and established a thriving business as *the* specialist in the design and construction of wooden velodromes for bicycle racing. Anyone who has witnessed a roller-skating derby on a banked wooden track has a good idea of what these much more steeply banked one-eighth and one-quarter mile ovals and saucers were like. Then Prince began building board tracks up to a half-mile in circumference for the new sport of motorcycle racing. And he and Moscovics crossed trails from time to time and kept their friendship alive.

After the Allen-Kingston episode (the company folded in 1909, a casualty of the previous year's economic depression) Moscovics' new employer sent him to Los Angeles, where he promptly made contact with the racing fraternity. There he met a well-to-do former racing driver named Frank A. Gurbut. They became good friends,

The Beverly Hills Board Speedway, seen from the air. TED WILSON

The Dusenberg team on the boards in 1919. Augie is bare-headed. On his left is Eddie Miller; below is Milton; squatting at right is Murphy. Note the huge splinters on track. GB COLLECTION

and when Fisher began to build his speedway in Indianapolis they agreed that the time was as ripe in Southern California as it was in the Midwest. But what sort of track should it be?

"A huge wooden saucer!" Moscovics announced, stunning his about-to-be partner. "Nothing can be as cheap, as fast, or as safe. And I know just the man who can build it for us."

The world's first one-mile board track was built at Playa del Rey, just a few miles from Los Angeles. Its first race was held on April 8, 1910, and was a smashing success in every way.

The track was an astonishing structure, the like of which never had been seen. It was 45 feet wide and was banked one foot for each three of its width: a very steeply curved, perfectly circular bowl. Gurbut had the idea of lining its upper edge with steel guard rails. They cost money, saved lives, and probably set the pattern for the highway guard rails that were to become almost universal decades later. The board track itself cost only 75,000 dollars to build, such were the costs of labor and materials. Then the farsighted owners invested another 10,000 dollars in a generating plant for lighting their wooden motordrome for night racing!

Top: Pillsbury's spiral easement curves, first used at Beverly Hills, permitted tremendous sustained speeds. Left center: east curve of the Beverly Hills track; note substantial guard rail. Right center: elegant covered grandstands at Beverly Hills, here under construction, added to the glamour and prestige of the speedway era. Right: Speedway racing attracted fantastic crowds; this was Culver City. A. C. PILLSBURY

Moscovics did much of the designing on the project and placed the lower guard rail 17.5 inches above the track level, to snub the hubs of cars running on 34- to 36-inch-diameter tires. A second rail was placed 12 inches higher to arrest tendencies of out-of-control cars to overturn. Then, between the track and the turf infield, the designers laid a 30-foot-wide strip of packed, crushed granite. Thus, it was almost impossible for a car to run off the outer edge of the track and, if it ran off the inner edge, it would have a hard time getting into trouble. The infield fence was located 125 feet from the track edge for the safety of infield spectators. Equal care was taken for the safety of spectators in the huge and comfortable covered grandstand. During its first week of operation the new plant pulled as many as 15,000 spectators in a day—fantastic in view of Los Angeles' cow-town population. Now, with a metropolitan population heading toward 5 million that is still a good crowd for a first-class event. The Los Angeles Motordrome gripped the attention of the whole American racing world, and it was hailed as the world's safest as well as fastest race course.

Atlanta, having a smaller and less prosperous population to draw upon, was a failure, and so was its gravel surface. Indianapolis went through the agonies of the damned learning that the best tar and crushed rock pavement of the period went to pieces immediately under the punishment of racing. In spite of crushing financial obstacles Fisher resurfaced his speedway with 3,200,000 paving bricks, weathered all storms, and created what was to become a world-renowned institution. But out in the Wild West was that wonder of a board speedway where absolute records tumbled at every meet until the day in 1913 when it burned to the ground with, according to Damon Runyan, "a great saving of life." But it had shown the way that was to govern and shape American racing for almost two decades, and decades of real glory they were.

While Playa del Rey was burning, Jack Prince was one of the world's busiest men, dashing from city to city and coast to coast, spreading the new gospel among promoters. He found not one taker for his talents but several, so that 1915 ushered in a nation-wide circuit of superb new board tracks. Maywood Speedway at Chicago was a two-mile oval, as were the tracks at Tacoma and Sheepshead Bay (Brooklyn). Of the total of twenty-four board tracks that existed between 1910 and 1931 these were the largest ever built and, had they survived, they still would rank among the wonders of the world. Also in 1915 a 1.25-mile board speedway was opened at Omaha and a one-mile saucer at Des Moines.

These extremely fast tracks consisted of two straightaways joined by 180-degree curves whose bankings were as steep as 45 degrees. The present Monza banking reaches only 38 degrees.

The Speedway Era formally began on June 26, 1915, with the inaugural event at Chicago, a 500-mile race. Indianapolis had drawn about 60,000 spectators three weeks before and the winning average of de Palma's Mercedes had been 89.84 MPH. Chicago drew about 80,000 and the winning average of Dario Resta's Peugeot was 97.58 MPH, a fabulous world's record on many counts. Journalists were at a loss for words to describe what had taken place.

Speedway racing, meaning board speedways, did much more than merely provide the world for a decade and a half with some of the most fantastic spectacles in history, automotive or otherwise. It was the board tracks, with their almost boundless possibilities for speed, which brought about the remarkably swift evolution of the American thoroughbred race car and which astonished the Europeans who tried to repeat their overseas conquests after World War I. It was the boards, which permitted full-throttle driving at all times, that dictated the quick adoption and brilliant development of the straight-eight engine, of superchanging, streamlining, and front-wheel drive. It was the boards that spurred brilliant progress in tire technology and in the development of fuels, alloys, bearings, and clutches. It was a period during which racing did improve the breed in every sense, including that of the passenger-car industry.

The Indianapolis "500" never lost its major status, thanks to a management which made it a point always to offer the world's biggest purse, thereby making secure its rank as the world's most important racing event. However, any car that could win on the boards could win on the much slower bricks, providing its chassis were strong enough to take their pounding. Indianapolis was one event per year, but during most of this epoch all the other Championship contests were fought out on the board speedways and always under the stress of much higher speeds. Hence their

Scenes like this were no rarity during the board-track era. At Cotati, Milton and Murphy pass the Elliott-McKee wreckage. EDDIE MILLER

fundamental importance in shaping the breed. Art Pillsbury knew them intimately. He built the best ones:

> Jack Prince was building all the board tracks and I was interested in racing. I was a civil engineer for the Rodeo Land and Water Company, which owned all of that big agricultural tract known as Beverly Hills. William Danziger, one of the company's directors, was also interested in racing. One day he came to me and said, "We're going to have the world's finest board track, and you're going to build it."
>
> Prince was the only man who had built board speedways up to that time so I naturally turned to him as a consultant. I made a deal with him: One, he would get a flat fee of $5000 for his services, including access to the designs of all his previous automotive tracks. Two, he would get credit as the designer of the Beverly Hills Speedway. Three, I would be free to try to engineer the new track since, as all of us knew, all the previous ones had been designed by guess and by God and according to the whims of the individual track owners. Prince agreed, accepted $500 for his travel expenses to come west and then collected his five grand.
>
> Then we began to get down to business and it became clear that there were no figures, no designs and that Prince was quite innocent of any engineering knowledge. He was very helpful of course but that first track that I designed and built in 1919 was the first one built with any serious attention to engineering principles.
>
> As vertical supports for the floor I used 2 x 12's on four-foot centers. Prince had gone to using 2 x 3 planks on edge for his floors but I went back to 2 x 4's and to straight, vertical-grain lumber to minimize splintering.
>
> One of the notorious defects of the earlier banked ovals had been the difficulty of getting on and off the curves. The transitions were abrupt, tricky and dangerous. So I used a Searles Spiral Easement Curve. This is a formula widely used in railroad engineering whereby a train is led into a central curve through a series of small curves of ever-decreasing radius. The formula includes the elevation of the outer rail to a calculated degree and the purpose of all this is to achieve a smooth ride around curves at speed. So, where I might have a radius for the main curve of 1000 feet, the first 30 feet going into it might have a 5000-foot radius. I changed the radius getting into the main curve, and leaving it, every 30 feet. Then I brought in a series of transverse vertical and parabolic curves.
>
> There was no limit to the speed for which those tracks could be designed. If you want one to handle 160 or 180 MPH it's just a question of how steep you throw it up in the air. There was no steering. You could just take your hands off the wheel. You could drive flat-out blindfolded. That's the way those tracks were designed.
>
> That first track was a private venture of Danziger's and ten other men, including Louis B. Mayer and Cliff Durant. They bought 200 acres from the

Rodeo Company, in the very heart of what became the city of Beverly Hills. I put the whole mile-and-a-quarter track up in five days. I had 1200 men on the job. The way I did it was to put up the understructure and then lay out the 12- to 18-foot 2x4's all around it. The lengths varied so that the joints could be staggered, as in bricklaying. Then I set up teams of two carpenters each, with two laborers to each team. The moment the first 2x4 was laid by the first team another followed until there were 300 of these teams working right around the oval. There was nothing to it—just manpower.

I finished the whole plant in five weeks. It was beautiful and we spared no expense. We had roofed grandstands with large boxes, each holding ten hand-built chairs that were contoured for real comfort. Everything was de luxe and so was our clientele, which included the whole film colony and just about everyone of importance within travelling distance of L. A. The attendance was tremendous and the plant made nothing but money for five years, when it was sold because it couldn't afford to stay there. Those 200 acres had been bought for $1000 each. They were sold for exactly ten times that.

Of course we wanted another track and a group of us bought the property in Culver City between the MGM Studios and what finally became Jefferson Boulevard. Beverly was sold early in '24 and, to hold our AAA franchise, we had to be in operation by Thanksgiving Day. We had to pay $4000 each for those 160 acres and we couldn't raise the necessary capital in time. When we opened we were over a quarter of a million in debt.

Then, we thought we'd save money and we left the roof off the grandstand. We cut the boxes down to eight seats and used folding chairs. We made the grand-

1924 action on the Ascot dirt track, Los Angeles. It was the training ground of many of America's finest drivers and mechanics. TED WILSON

stand seats out of just flat planks. We cheapened the plant and we didn't get the same clientele that came to Beverly.

You can't open a speedway plant of any kind and apologize for it. You can't say, "We'll have it better for you a year from now." You've got to have all the facilities for free ingress and egress and you can't apologize for that.

Well, when we bought this property we were promised that Jefferson Boulevard would be in service before we opened. It never was and from the first race we were five to six hours getting the people off the plant. We could get a hundred thousand people in there but we couldn't get them out because there was only one exit. That gave us a black eye from our opening day and in four years the bank that had our loan foreclosed on us.

So that's why Culver failed. But what happened to all the other board tracks? For one thing, in those days we knew of no preservative that we could use to treat the lumber to prolong its life that did not become slippery under friction. All we knew was creosote. We tried everything, but nothing worked. So, at the end of about four or five years, depending upon climatic conditions, the life of the 2x4's, although they were laid on edge, became sapped out. They became brittle and the floors started going to pieces. No floor was good for more than seven and a half years.

The floor was the real cost of the structure. It took a million board feet for a mile and a quarter track—about $125,000 in those days and well over double that today. With modern technology there is probably some way that you could put this lumber in tanks and pressurize it and perhaps extend its life almost indefinitely.

Lockhart lapping the Atlantic City boards at 147.7 MPH in 1927. TED WILSON

The cost of the floor could have been overcome in any case. The real and final failure was that, as the cars became more perfect, any real racing ceased on the boards. The fastest car never was headed. There was no driving involved, there were just so many squirrels going around in a cage and positions changed only when there was some mechanical failure. Racing was lost and that, in the end, was probably the deciding factor.

Pillsbury's scientific approach to the design of board speedways resulted in such success at Beverly Hills and in such tremendously elevated speeds that he was called upon to design and direct the building of half of the fourteen board tracks which followed. During most of the same period he served as Western Zone Supervisor for the AAA Contest Board, which he had done much to create. He prospered from his diverse enterprises. He ruled the sport by iron law. He was often unpopular because of this, but his fierce integrity never could be questioned. During the Great Depression, which helped to finish the golden era, it was Pillsbury who, out of his own personal capital, did much to underwrite and make possible the survival of first-rate racing in the western United States which, by then, had become its home. He played a decisive role in the development of the Bonneville Salt Flats as a speed course. He masterminded the original Gilmore Economy Runs, which evolved into the Mobilgas Economy Runs, all of which have been subject to his stern law and obsession with impartiality. The last track he designed was Roosevelt International Raceway on Long Island in 1936.

Building the boards was one thing, driving them something else. Eddie Miller drove them for Duesenberg:

Sheepshead Bay! What a beautiful thing it was, and two miles around! If we had it today we could lap it at an easy 180, maybe 200. We ran all those terribly fast tracks with wonderful reliability. I loved that kind of racing until the tracks began falling apart.

You'd show up for a race, a few holes would show up in the floor during practice but the carpenters would patch them up and things would look pretty good. Then, when you had maybe fifteen cars starting a 250-mile race, holes would start developing under the pounding of the machines. They would have carpenters working away under the track. When a hole started just as a three or four-inch crack there was nothing to do about it until it got bigger. Sometimes a crack would develop as much as a foot wide. Then you'd see a carpenter's head stick up. He'd nail something on the side and duck his head down, then come back up and nail another one on. But he never could get the hole plugged. And kids who sneaked into the tracks used to watch the races from those holes, and between the kids and the carpenters, all those human heads bobbing up and down didn't exactly make for the most serene of driving conditions. But they made more than one driver swear off of drink for a day or two.

I remember times at Uniontown where there would be as many as ten of these openings around the track, all from a foot to a foot and a half wide and

Leon Duray's 124.7 MPH for 250 miles was an absolute world record. So it was claimed, was mechanic Dick Doyle's eight-second tire change. EDITH SAVAGE DURAY

up to about forty feet long. All the way around you had either to dodge those things or to straddle them. That was hell at 110 MPH and more and a lot of the boys got caught and stacked up.

You used to get hit with some terrific blocks and knots of wood. We all came in with pieces of wood bigger than kitchen matches driven into our faces and foreheads. They'd go in, hit the bone and spread out. Then you had to remove them, of course. Tacoma was worse. You had the splinters and knots and all, but to save on lumber they had spaced out the 2x4's and caulked them with some mixture of tar and crushed rock. When Tacoma began to go it was like a meteor shower.

At those speeds in those days we had our problems with tires. The casings were remarkably good but we all lived in terror of throwing treads. At 100 MPH or so, half a tread would come flinging off, throwing the wheel violently out of balance. Well, you'd go hopping along hoping that the rest would fly off, particularly if you didn't have too far to go to finish the race. We finished plenty riding just on the cords. Well, when you're riding on a board speedway, with those planks polished just like a dance floor and all you have under you is those shiny cords, I'm telling you, Mister, that it was a pretty delicate job keeping that car on the track, not to mention dodging the holes.

The ephemeral, forgotten boards were America's real crucible of speed, and the men who drove them were a breed indeed.

chapter 4

THE SOURCES

AT THE TURN OF THE CENTURY racing cars in the United States were almost exclusively ordinary production cars. Toward the middle of the first decade cars designed specifically for racing began to appear, but by the end of the decade the trend had swung back to the stock car, cut down and otherwise modified for increased performance. Ray Harroun's Marmon *Wasp,* winner of the first Indianapolis "500" in 1911, is a typical example. There were occasional exceptions to this pattern but it was dominant until 1913, when the Europeans arrived at Indianapolis and put an end to the modified production car as a serious racing vehicle.

Some quite amazing performances were achieved by pioneer American one-off race cars. In 1903 R. E. Olds built his tiny, spidery, superlight *Pirate,* which weighed only 825 pounds and was clocked at 53.5 MPH at Daytona that year. On the same course in 1904 Alexander Winton drove his straight-eight *Bullet No. 2* past the official timers at 92.4 MPH. In 1905 Louis Ross in a steam car with a remarkably streamlined body pushed the record for the flying mile to 94.73. And the following year Frank Marriott, in an equally streamlined Stanley steamer, set a new world's record for speed on land at the incredible—for 1906—velocity of 127.6 MPH.

The first man to break the mile-a-minute barrier on a one-mile dirt track was Barney Oldfield, fresh from bicycle racing on the velodromes. It was in 1903 in a race at Indianapolis which had been organized by young Carl Fisher, and the car was the *Red Devil,* twin to the famous *999,* which Henry Ford had built the previous year.

In 1903 Barney Oldfield became the first man to break the mile-a-minute barrier on a mile dirt track. The car, *999*, was designed and built by Henry Ford. Note the exposed overhead camshaft. GB COLLECTION

The following year young Ford himself fought the monster's tiller steering over an iced-over lake to a new absolute "land" speed record of 91.37 MPH.

There were some interesting features in the *Red Devil's* 7 by 7 inch bore and stroke engine. Its four cylinders were cast *en bloc* in anticipation of what would become general practice. Its inlet valves were of the then common automatic type, piston suction pulling them open against their weak springs. The exhaust valves, however, were operated by an overhead camshaft which was driven by a not-quite vertical shaft and bevel gears. While there is rarely anything really new under the sun this certainly was an early use of this principle. The whole valve train ran in the open air, directly in front of the totally exposed driver, and the crankcase, so-called, was just an open latticework of supports. The heroic driver of this heroic machine was drenched with oil from the moment the engine was started, and Ford himself stated, "The roar of those cylinders alone was enough to half kill a man."

Ford actually built two of these cars. The one which survives today in the Henry Ford Museum has passed through many hands and all seem to have made their improvements, including deletion of the SOHC. But it looms, in all its advanced glory, in old photographs of the car.

The earliest recorded use of vee-inclined overhead valves in a European engine is the Belgian Pipe of 1906, which also had its spark plugs located in the center of the cylinder head. Its crankcase followed conventional T-head practice, with a

In 1904 Ford himself (with goggles) set a record of 91.37 MPH in the same car.
Autocar, LONDON

camshaft slightly above and on each side of the crankshaft, then with pushrods to operate the widely inclined overhead valves. The Pipe was a milestone in engine design.

Strangely enough, although Carl Fisher's one-off 1905 Premier has been seen by hundreds of thousands of visitors to the Indianapolis Speedway Museum since its opening in 1956, this milestone has been totally overlooked. This is remarkable, since it appears to possess the world's first engine in which the principles of overhead camshaft and inclined overhead valves were combined. It was air-cooled and was one of the first American cars to use magneto ignition. It was a most prophetic machine.

Carl Fisher wanted a potential winner to drive in the last of the Gordon Bennett Cup Races, to be held that year in Auvergne, France, and placed his order with the Premier Motor Corporation of Indianapolis. Co-owner and chief engineer of the firm, George A. Weidely, undertook this command performance. He had created a very successful small, air-cooled four cylinder engine which he mounted transversely, Issigonis-style, in 1903, at the front of a utility vehicle, a considerable number of which were manufactured and sold. He then drew up crankcases which permitted two of these 3.75 by 4.25 pushrod OHV engines to be bolted together in line. This pioneer straight eight, quite an immaculate piece of design and workmanship, was called the Premier *Comet* and was raced by Fisher with great success, which is why Fisher turned to Premier when he wanted a challenger for what was, then, the most important speed contest in the world.

The engine which Weidely designed had a drastically over-square bore and stroke of 7 by 5.5 inches, for a total displacement of 846.6 cubic inches. The shaft-and-bevel SOHC operated the valves (included angle 45 degrees) by means of rocker arms whose leverage ratios were scientifically calculatd. Moreover, the designer was articulately and precisely aware of the advantages of this entire configuration, all of which added up to superior efficiency.

What got to be overlooked in the heat of creation was the one limitation in the rules for all the Gordon Bennett Cup Races: competing vehicles might not weigh more than 1000 kilograms, 2200 pounds. The huge new Premier exceeded this limit by many hundreds of pounds, and all the hole-drilling that was possible could not trim its weight sufficiently. Fisher drove it in just one race—at the Indianapolis Fairgrounds horse track on November 4, 1905—and, though saddled with a 78-second handicap, he won the contest, turning the final lap at 59.21 MPH. Then the car, which had cost Fisher 15,000 dollars, raced no more. Why it did not is hinted at in the words of journalist Hugh Dolnar, who rode in the machine on one of its first shakedown runs:

> A twelve-mile round trip on East Washington Street with trolleys in the middle and horse-drawn vehicles on the side, was not a tranquil joy to this

Left: The remarkably advanced 1905 Premier engine in transverse cross-section. Right: In 1905 George Weidely of the Premier Motor Corporation designed this air-cooled, SOHC, hemi-head racing engine. GB COLLECTION

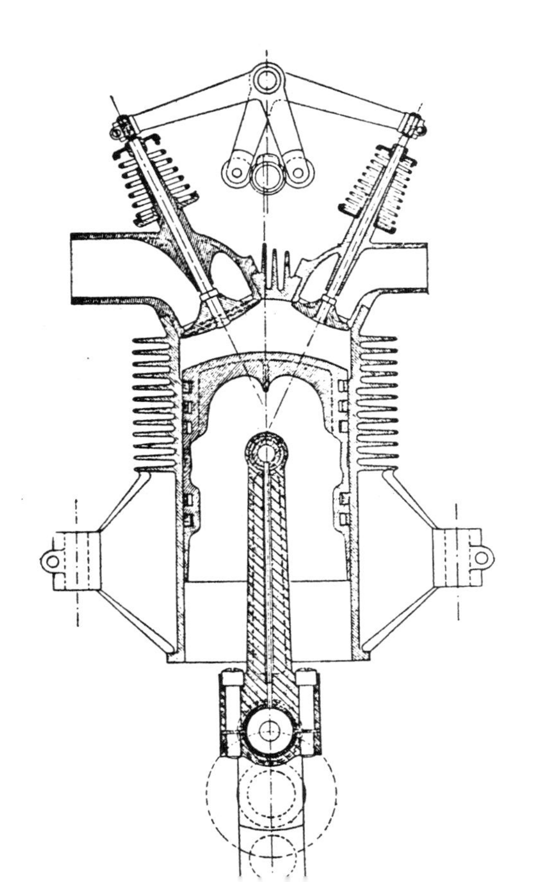

Winner of first "500" was Ray Harroun's Marmon *Wasp,* based on touring chassis. IMS

> writer. Mr. Weidely says that the seats on this racer are remarkably easy riding. This may be true in a comparative sense, or on a smooth track; to make it unreservedly true the car must be placed on a far better surface than the East Washington asphalt, which is not so bad, but is not good enough to drive a mile a minute on with comfort. This car can run fast, perhaps a mile in 40 seconds. It ran at about a mile a minute clip for a short distance on East Washington Street with the throttle very slightly opened, and the exhaust muffled to a mild roar, not much louder than Niagara Falls and, say, a thousand threshing machines in active operation combined. What this racer will do with no muffler and the throttle wide open the writer does not know, and does not want to know, and never will know, of his own knowledge.

George Weidely is, lamentably, an unsung hero of technological history. He sensed the way things should be to a startling extent. In the Fisher car he used shaft drive, while most others would continue using chains for years (chain drive was substituted in the desperate efforts to reduce the car's weight). He used a multiple-disc clutch while others stuck with the leather-faced cone type for even more years. Weidely was far in advance of his time, but the merit of his ideas was lost on the rest of the world.

It is not being chauvinistic but merely factual to note that "Yankee ingenuity" soon came to dominate most areas of world technology. It was more than any mere knack for things mechanical or technical; it was a spirit of openness, hope, and opportunity which provided the environment in which this ingenuity could flower. Take Henry M. Leland, for example, one of the greatest of technological geniuses, whose total inspiration came from his devout belief in the blessings that technology could bring humanity. He founded and served as chief engineer of the Cadillac Motor Car Company, made a gift to the world of the techniques for the mass production of precise interchangeable parts, gave up his interest and position in General Motors because of its side-line posture in World War I, and founded the Lincoln Motor Company to serve the Allied war effort, because he believed it to be right. It was his lifelong pride that at the age of 21 he had cast his first presidential vote for Abraham Lincoln. On the lawn before the Lincoln plant in Detroit he placed a life-sized statue of the liberator, and on its pedestal had engraved his words, "Let Men Be Free." It is still there and, in spite of some obstacles, so is the basic spirit that underlies all the real greatness that the country has achieved or ever will achieve, aided by its bounteous resources and hard-won freedom from the shackles of countless outworn and suffocating traditions.

The mechanical arts in the United States had begun to catch up with and occasionally surpass those of Europe when the basic configuration of the classic racing engine finally was resolved in the prewar Grand Prix Peugeots. Our man in Europe, W. F. Bradley, witnessed the birth.

> The old firm of *Les Fils de Peugeot Frères* was active in *voiturette* or light-car racing; it called its competition cars Lyon Peugeots to avoid having them confused with the very dissimilar Peugeot production cars. This class of racing was dominated by this marque and by Sizaire-Naudin until, in 1910, engineer Marc Birkigt turned loose a new four-cylinder, T-head Hispano-Suiza in the *Coupe des Voiturettes* at Boulogne in France. The Italian-born driver for the Spanish factory, Paolo Zuccarelli, won the race decisively against all opposition, including that of the Lyon Peugeot aces, Georges Boillot and Jules Goux. A result of this encounter was the development of a strong friendship, based on respect for each other's talents, between these three men.

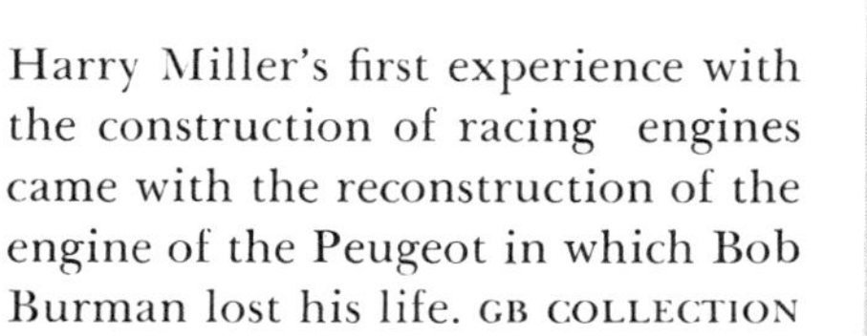

Harry Miller's first experience with the construction of racing engines came with the reconstruction of the engine of the Peugeot in which Bob Burman lost his life. GB COLLECTION

They all were bored with driving little cars and wanted to be in racing on the highest level. Out of their conversations came the idea that they, mere drivers, knew more about how a race car should be built and how it should function than any textbook engineer possibly could. Agreeing on this, they proceeded to put on paper their conception of the ideal Grand Prix machine.

Goux' entire family worked for Peugeot, as he had since he was a boy. He was the personal chauffeur of Robert Peugeot himself. He was a very serious young man who never answered a question without thinking it over several times and he was held in the very highest regard by his employers. Therefore he was chosen to present the idea to them. Without the confidence in which he was held the dream would have died right there.

Goux made his sober presentation and Robert Peugeot said, "Jules, go ahead. We will back you. We have a little factory just outside Paris that you can make use of."

When this news reached the firm's proud, professional engineering staff there was a near rebellion. They scoffed at Peugeot's folly and referred to the three upstart *mere* drivers as *Les Charlatans.*

But the charlatans were a gifted team. Zuccarelli was responsible for perhaps 80 per cent of the ideas, many of which he had absorbed working closely with Birkigt in Barcelona and Piccard-Pictet in Geneva.

The trio had roughed out the general design of the original Grand Prix Peugeot and needed a draftsman to put it properly on paper and detail it. They hired the former Piccard-Pictet draftsman, Ernest Henry. Many English writers cannot imagine that there could have been a successful race team without an

The 300-mile Indianapolis race of 1916 was won by Resta in this Peugeot at an average of 83.26 MPH. He qualified at 94.47. IMS

engineer, and thus they have invented the Henry School. In Europe a man must have a diploma to be an engineer. Henry had none; nor did any of the other members of the team. He was a very modest chap and never claimed to be anything more than a draftsman. His great feature was that he was very quick to seize ideas, put them on paper and make them work. He was not inventive and never introduced any new ideas of his own.

The first car was built in 1911. No part of any of these cars was built in the Peugeot factories; everything was farmed out to little shops around Paris. The project was the laughing stock of the city but Robert Peugeot never wavered in keeping his word. Boillot drove the first car in its first contest—nothing less than the Grand Prix of the Automobile Club of France in 1912 at Dieppe. He ran away from Wagner on the all-conquering Fiat. The Peugeot had 7.6 litres against Fiat's 16 litres and the race was for a distance of 956 miles. Peugeot's name was made overnight in big-time racing and thereafter it was a rare race, hill-climb or reliability test that these cars entered and did not win.

Boillot won the French GP again in 1913 and Goux came in second; their cars outclassed all the rest. It was at that point that Sedwick arrived, seeking European entries for Indianapolis and, above all, the GP Peugeots. Boillot let two of the old cars go, and Goux won the race. They went again in 1914 and lost to Delage because of inadequate tires. I was the manager of the Delage team and I know that the Peugeots were faster cars and that, with better tires, they would have won the race.

Then there was the last French Grand Prix, less than four weeks before the outbreak of the war. *Les Charlatans* had begun to suffer from lack of harmony in their little group and the Mercedes were definitely faster on the long straightaway at Lyons, although slower in the switchbacks. In pursuit of the German cars Boillot seized a piston on the final lap while Goux, driving with his usual conservative strategy, came in fourth behind the three victorious Mercedes. And that was the end of the wonderful Peugeot racing team which, in my mind, is the best that the world has ever seen. Just three drivers and a draftsman.

Three drivers and a draftsman wrought the great and basic revolution in racing car design and, almost overnight, the whole racing world was copying their ideas. To their despair the Peugeot management made it a policy to sell the GP cars as soon as they had run a race or two and a buyer came forward. New and better cars could always be built, they seemed to feel. They had many more pressing concerns than Grand Prix racing, being the proprietors of a vast and diversified industrial complex. They had probably backed the project in the first place only as a lark and as a patronizing gesture of good will toward the dutifully serious and hard-working Goux family. *Les Charlatans* saw their secrets being broadcast to the entire world and they became demoralized. Then the war scattered them.

By 1914 copies of Grand Prix Peugeots had showed up in nearly every European country which built cars. Louis Coatelen, the French chief engineer of Sunbeam in

England, had a French associate buy one of these machines, took it apart, copied every piece, put it back together again and rented it out at so much per week for other British engineers to study. Humber, Straker-Squire, and Vauxhall were among the beneficiaries of this program. Delage in France, Opel in Germany, Nagant in Belgium, and Aquila Italiana in Italy also built copies of the GP Peugeots.

Across the Atlantic it was the same; Peugeot was the car to own or copy. Several GP Peugeots were sold to individuals in the States, including Dario Resta and Bob Burman, whose broken engine and chassis were the textbooks for Harry Miller and Fred Offenhauser. Then, with the outbreak of war in Europe, the Indianapolis management became concerned about losing the crowd appeal of "international" participation in its events and ordered several Peugeot cars from France. The factory replied that it was out of the racing business and totally involved in the war effort. The Speedway therefore commissioned the Premier Company to build three replicas of the GP Peugeot. These copies, owned and campaigned by the Speedway and which raced under the names of Peugeot, Premier, and Premier-Peugeot, were quite successful. They were such accurate copies that, according to legend, when Goux came over in 1919 with one of the prewar cars and cracked a cylinder block in practice the day before the "500," a Premier-Peugeot block was sent for, dropped

The 1913 Grand-Prix Peugeot set the classic pattern and was the most widely copied in racing history. W.F. BRADLEY

When Peugeot abandoned racing in 1914 the Indianapolis Speedway management commissioned Premier to build these close replicas of the GP Peugeot. They began racing in 1916. IMS-GEBBY

directly onto the studs of his crankcase, and enabled Goux to finish third in the big race. It seems to be both ironic and appropriate that George Weidely's organization built these replicas, having anticipated the originals to such a striking degree.

Perhaps every design feature of the 1913 GP Peugeot had been used before, but never in this harmonious and overwhelmingly successful combination. The full technical details are to be found in Volume I of Laurence Pomeroy's *The Grand Prix Car,* and W. F. Bradley tells the human story in his *Motor Racing Memories.*

The following elements were integrated by *Les Charlatans* in the original GP Peugeot. Its four cylinders were cast as a single iron block with integral head. The four overhead valves per cylinder were inclined to form an included angle of 45 degrees. They were operated by dual overhead camshafts which were driven by a vertical shaft and bevel gears. The spark plugs were located centrally in the pent-roof combustion chambers. The crankcase was of aluminum, split down its horizontal center-line, and the crankshaft ran in five plain main bearings. A Robert Bosch magneto provided current for ignition. A crowning touch was the use of desmodromic valve actuation. Bore and stroke were 110 by 200 Millimeter (4.33 by 7.88 inches).

The smaller, 78 by 156 Millimeter (3.07 by 6.14 inches), GP Peugeot of 1913 was a much refined version of the original concept. The dual overhead camshafts, four valves per cylinder and integral-head block were retained. The included angle between the valves was increased to 60 degrees and the desmodromic, "mechanical" closing of the valves gave way to cup or piston-type cam followers—the earliest known use of this principle, although the idea undoubtedly was not new. The shaft-and-bevel drive to the camshafts was replaced by a much more accurate and manageable train of spur gears in their own compact, quickly detachable aluminum housing. The split crank-case was replaced by one of the barrel type in which the crankshaft, with its main bearings, is inserted from the flywheel end of the case. The crankshaft was counterbalanced and the five plain main bearings were replaced by three large ball races. Ball-bearing technology still was primitive and pressure oil was fed with the aid of centrifugal force to the main bearings through grooves, covered by steel bands, in the circular crank webs. And dry-sump lubrication was adopted, this being one of the first applications of the principle. This engine was built in 5.65- and 3-liter (344- and 183-cubic-inch) versions, and the smaller the GP Peugeots became the faster they eventually went.

This was the classical configuration that almost the entire racing world copied immediately, recognizing its superiority, or reluctantly copied at a later time in order to remain in contention. With the exception of the single-cam Duesenbergs, all American racing engines that have enjoyed any success since the end of World War I can trace their lineage directly to the 1913 creation, the work of "just three drivers and a draftsman."

When they separated at the start of the war Henry went to work for Bara of Levallois, near Paris, where the Bugatti "twin straight eight" was manufactured. During the war René Thomas dreamed of repeating his 1914 Indianapolis victory on a Delage, just ahead of Arthur Duray's Peugeot. When peace began to loom he went to his friend, engine-manufacturer Ernest Ballot, and asked if he would like to win the Indianapolis "500." Henry had nothing but time on his hands at Bara and could draw up a winning car. Thomas had brought home 28,000 dollars in 1914, and the postwar purse should be bigger. Ballot agreed.

With some counsel from Thomas and some Bugatti ideas, but mainly drawing

A similar stable of thoroughbreds was built under the auspices of Carl Fisher's Prest-O-Lite Corporation. These four-cylinder SOHC Maxwells were remarkably fast. JAMES TALMADGE

on his background with *Les Charlatans,* Henry went to work on what was essentially a straight-eight version of the 1913 GP Peugeot. Ballot had nothing to do with the design but got the credit; Henry designed the entire car and remained almost anonymous. As soon as the war ended four machines went into production and, in spite of the catastrophic conditions of the time, all four were running under their own power at the end of 102 days. They were beautiful.

The Ballot probably could have swept the field at Indianapolis in 1919. They were proved the fastest cars when Thomas set a new qualifying record of 104.70 MPH. However, Henry, perhaps wanting to pamper his engines, insisted upon a very high axle ratio for the cars over the protests of Thomas, who knew the track, which Henry had never seen. In practice it was immediately clear that Thomas had been correct. There was no time to have new gears cut, so Thomas replaced the large French Rudge-Whitworth wheels with smaller American ones, and this change in effective gearing made possible his record, plus teammate Louis Wagner's new record of 101.70—the third fastest time after Louis Chevrolet's Frontenac at 103.10.

But the American wheels were weak, and one of Wagner's broke at the hub on Lap 44, sending him crashing into the wall. Paul Bablot had been relieved by Jean Chassagne, and on Lap 63 one of his wheels collapsed and his car overturned and was wrecked. It is fortunate that no serious injuries resulted from these disasters, particularly since Chassagne's mécanicien was a young Frenchman named Jean Marcenac, who was to become one of the great talents in American racing history. It took courage after these grim warnings for Thomas and Albert Guyot to remain in the race, but they reduced speed slightly and made frequent stops to check their wheels. They finished tenth and fourth, respectively.

Because of bad luck the Ballot campaign was abortive, but it made a profound impression. It made it clear that the straight eight was the engine of the future. In doing so, it confirmed Fred Duesenberg in having already arrived at the same conclusion and in having committed his own future to it.

The Ballot team returned in 1920 with even finer, brand-new 183-cubic-inch-engined cars. This time the gear ratios were right, the cars were perfect, and both Ballot and Henry came with the team for the first time to insure and witness a sweeping victory which was beyond any doubt. This time the Ballots' excellence was even more marked, de Palma being fastest in qualifying at 99.15 against second-place Joe Boyer's Frontenac at 96.90. De Palma was equally outstanding in the race, at one time having a full two laps over his nearest competition. But the race went to Gaston Chevrolet's Monroe, with the Ballots of Thomas, de Palma, and Chassagne finishing second, fifth, and seventh. W. F. Bradley recalled:

> I was there in 1920, when I headed the Gregoire team, the worst ever to cross the Atlantic.
>
> Chevrolet showed his superiority by winning, but I do not think, however, that he had quite obtained the leadership. Having nothing to do, I was watching from the Ballot pits. With only about three laps to go, de Palma was a certain winner

The first European cars designed specifically for speedway racing were Ernest Henry's Ballots. In 1921 de Palma won a 50-mile race on the Beverly Hills boards at 107.3 MPH. GB COLLECTION

and Ballot began to write a cable to Paris announcing this, with only a gap for the winning time. Then the engine spluttered. It was thought to be a lack of gas, so the car was refuelled. But the spluttering continued and Henry said it was ignition, that one of the two four-cylinder magnetos had failed. Ralph had such a lead and there was so little time left that Henry told Ralph's mechanic, de Paolo, to remove the four non-firing plugs so as not to squander compression pressure. They finished the race on four cylinders and of course were overtaken by Chevrolet. Thomas was given the hurry-up call but was too far back to be of any use. Afterwards it was discovered that a cork gasket had disintegrated in the fuel system of Ralph's car and this was the cause of the misfiring.

We in Europe never realized that while we were at war, America was continuing much as usual; that Indianapolis was closed but that many of the great board speedways continued in full swing. The Americans had all the European cars to dissect—Peugeot, Delage, Sunbeam, Mercedes and others—and their technology was making strides that we could not dream of. I think that that was the feature of 1919–1920. American designers and drivers were becoming unbeatable on their own ground, and we reeled in astonishment at the progress that had been made in America while we had been busy fighting one another in Europe.

PART TWO

Masters of the Machine

chapter 5

FREDERIC SAMUEL DUESENBERG

FRED DUESENBERG WAS born in Lippe, Germany, on December 6, 1874. His brother August Samuel was born there in 1879. Their father died while they were very young, and their older brother Henry went to the United States to seek his fortune. He got a job as salesman for a large nursery company and was assigned to the then quite primitive territory of northeastern Iowa and adjoining Wisconsin. Rockford, Iowa, was in the approximate center of his sphere of duty, and there he settled down and sent for his mother and her other five children.

None of them had much in the way of formal education, but Fred and Augie both were blessed with unusual mechanical talent. At seventeen Fred was making his own living and helping to support his family by repairing farm machinery and erecting windmills.

The bicycle was the great, revolutionary form of personalized mechanical transport of the day, and Fred gravitated to it as rider, then as mechanic, then as racer, and then as builder of his own fast machines. At age twenty-one he had his own small manufacturing business, continued racing and, in 1898, was credited with having established the world's records for two and three miles.

Two years later he had learned enough about internal combustion engines to bolt one onto one of his bikes and become a pioneer motorcyclist, as Harry Miller had done earlier in nearby Wisconsin. Then, in 1902, he went to work for the Rambler Motor Car Company in Kenosha, Wisconsin, where he learned many of the

Fred Duesenberg at the wheel of one of his early Mason race cars, about 1911.
JERRY GEBBY

basic facts about automobiles. The following year he returned to Des Moines, where he organized the Iowa Automobile & Supply Company. Augie joined him and to promote the little enterprise they began modifying production cars and racing them and doing well. They became local celebrities.

It was just a matter of time until they built a car of their own from the ground up. It was designed to cope with the rugged road conditions of the rural Midwest, and an enterprising attorney named Mason recognized a ready market for such a car. He provided the capital and managerial acumen, Fred and Augie contributed their technical talents, which were to become legendary, and the Mason Motor Company was formed in 1907. The Mason car was soundly designed and built and, although its simple engine had only two cylinders, it plugged away like a faithful plowhorse over the rolling, Elysian Iowa countryside. The car was so good and sold so well that in 1910 the entire business was bought up by washing-machine tycoon Frank Maytag. Maytag Mason cars remained in production until 1916.

From the beginning, Mason had approved of racing as a means of promoting his product, and he encouraged his young "chief engineer" to undertake the development of an engine for this purpose. It was quaintly conceived and executed but the conception was an excellent one, and it soon began bringing fame to the Mason name on race tracks in far-flung parts of the country. Fred wisely retained the rights to this engine design, and when the Maytag purchase took place and the Mason plant was moved to Waterloo, Iowa, Fred went into business for himself, building race cars that he first campaigned as Masons and then gave his own name to.

Their success and remarkable efficiency led to many government contracts for marine and aircraft engines and then to his basic "walking-beam" power plant being adopted by many makes of high-quality, high-performance car. Fred's aircraft experience led to his pioneering, on this side of the Atlantic, the straight-eight engine configuration which immediately set the trend that the sport and industry were to follow for the next two decades. It also promptly enabled Duesenberg to break the world's land speed record and then to achieve America's first and, to date, only victory in a major European *grand épreuve*. That win of Jimmy Murphy's in the French Grand Prix of 1921 catapulted the Duesenberg name into world prominence and seemed to insure the success of the Model A Duesenberg passenger car, which was essentially a touring version of the thoroughbred racing machine. In its day it was one of the very finest cars in the world, but the management of the company was in the hands of Fred's financial backers. The company failed and was acquired, along with the talents of Fred and Augie, by young Erret Lobban Cord, who developed his Auburn-Cord-Duesenberg combine into one of America's richest financial empires.

In spite of their responsibilities with the design and production of the Model A and J production cars, Fred and Augie remained in the forefront of racing throughout the golden era, always inseparable. Then, with the death of the 91-cubic-inch formula, the brothers went separate ways, but only in a technical sense. Fred elected to base his racing activities on production-car engines—his own, of course—while Augie rented a shop across the street and went on building purebred race cars.

On July 26, 1932, Fred died. His Model J went into a skid on Mount Ligonier in Pennsylvania, and he was taken to a hospital in nearby Johnstown. He had a broken collarbone and ribs, a twisted spine, and other internal injuries, but it was ether pneumonia that took his life.

Duesenberg's racing brilliance was at an end, too, and know-nothings said that, with the guiding genius gone, the younger brother was incapable of accomplishing anything on his own. The reality was not that simple, nor was it like that at all. Tommy Milton said:

> I won't agree for a second that Fred was the strong man of the pair. He got all the glory and all the credit. But I think that Augie's contribution was at least as great as Fred's and certainly the whole Duesenberg effort never would have jelled without Augie. He did all the work. Fred did the promoting and had more to do with designing. But Augie was very capable in that field too and his thinking tended to be more practical. He was the sort of person who could make or

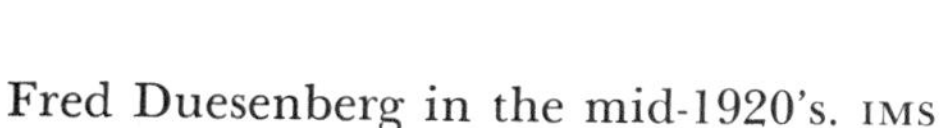

Fred Duesenberg in the mid-1920's. IMS

Augie Duesenberg in the mid-1920's. IMS

do anything and he worked tirelessly and endlessly. It obviously was a team operation and in my opinion Augie played the part to Fred that others played to Miller. Fred had the dreams and Augie made them work. But it was the old European idea of the older brother being the kingpin and the younger taking whatever happens to be left over. Both of them were out of racing and fed up with its heartbreaks before Fred died.

Said long-time Duesenberg mechanic Ernie Olson, who rode with Jimmy Murphy at Le Mans:

Fred was the sort of man that you wouldn't work just thirty-six or forty-eight hours for without stopping. You'd work an eager seventy-two. You couldn't do enough for him and he couldn't do enough for you. We worked for him like football rookies trying to please a great coach. Anything he told you was precise and you never had to guess. We worked hours that you wouldn't believe. Fred was always around until about two in the morning but Augie, being the younger brother, had to stay a little longer. It was a standing joke around the place that Fred would say at some God-forsaken hour, "Well, what do you say we all knock off and get a good night's sleep?" The ironic humor of it was that we all knew that we had to be back on the job and working like beavers at eight in the morning, *that* morning.

Augie was a top-notch mechanic and a fantastic welder. Fred was more the engineer. Neither was a businessman and the sharks of the financial world just tore them to pieces. They were both modest men, without pretensions. Augie worked with the rest of us on the cars at Le Mans and the Europeans never got used to the sight of *Mister* Duesenberg getting his hands dirty with the tools.

A constant refrain in the memories of nearly all who knew Fred is the genuine affection which he projected toward the whole world and which inspired real love for him in so many of the people around him. He did his best to live by a simple motto, "To be happy. To make others happy."

Another everlasting refrain is, "He was like a daddy to me." And, "We were one big family." Every member of the family was expected to work hard, but none worked harder than Fred and Augie. The only reason that, in the late Teens, Fred began knocking off an hour or two earlier than the rest of "the family" was that he was so crippled with arthritis that it was all he could do to hold a pencil.

In *Duesenberg,* by J. L. Elbert (Dan Post Publications, Arcadia, California), Pete de Paolo recalls about Fred:

> His ability as an outstanding engineer is world famous and is written in the records. No one can deny he qualified in the ranks of his profession as a true genius. With always a gesture blended with a tinge of comedy, he would express the outcome of the cars that he had designed and built, particularly with racing cars which were generally ready to qualify at the eleventh hour, by saying, "She'll run, kiddo," and they did. . . .

Fred and Augie may often have been slap-dash in their methods and eternally hounded by problems of time and money. However, Fred was the sort of man who when asked how he thought a part should be dimensioned, could then pick up a steel scale, gaze at it reflectively for a long moment, and then rattle off all the figures to the nearest sixty-fourth of an inch. Textbook engineers were appalled. They would make their own detailed calculations, and usually find that what Fred had sensed intuitively was very close to the ideal theoretical set of specifications for the part. Harry Miller had this same knack, but to a much less pronounced degree.

In 1927 a strangely epitaphlike bronze plaque was presented to Fred by the AAA Contest Board. Val Haresnape probably wrote the words:

> Racing is the crucible in which have been thoroughly tested many of the fundamentals of automotive engineering found in present day automobiles. The race track has been the stockroom of ideas for engineers of passenger cars, to which you have so graciously surrendered the keys.

Quiet, shy Augie worked for Auburn-Cord-Duesenberg during the Thirties, built the record-shattering *Mormon Meteors* for Ab Jenkins, and retained an intense interest in the sport until the age of 76 when, on January 18, 1955, he passed on. Both brothers are buried in Indianapolis.

chapter 6

LOUIS JOSEPH CHEVROLET

IF YOU HAD HELPED to create the largest-selling automobile in history and had even given it your own unusual name you might not expect to die in meager obscurity, and in Detroit at that. Louis Chevrolet did.

He was big and kind and a tender and loving family man, but if pushed far enough his wrath was fearsome and so was the bearlike strength with which he could vent it. He was one of America's best racing drivers, and he had a rare instinct for what was mechanically right. He worked and fought hard for everything he ever got out of life, such as it was, and he left huge tracks in history. But of all the men who have devoted their lives to the passion of the machine, none have reaped more miserable rewards.

Louis Joseph Chevrolet was born on Christmas day, 1878, in La Chaux de Fonds in the Bernese Jura of Switzerland. His father, Joseph Félicien Chevrolet, was a watch and clock maker, and he taught his sons about machinery and how to do fine work with their hands. In 1888 the family moved to Beaune, a small town in Burgundy, France. Louis' brother Arthur was born there in 1886, followed by Gaston in 1896.

Louis had scant formal education, and while he was a young boy he went to work as a guide for a blind wine merchant. He had an intuitive flair for things mechanical, and in his teens he designed and produced a wine pump which brought him local renown. From this the young entrepreneur progressed to manufacturing, racing, and marketing his own bicycles. He named them Frontenac, for the Seven-

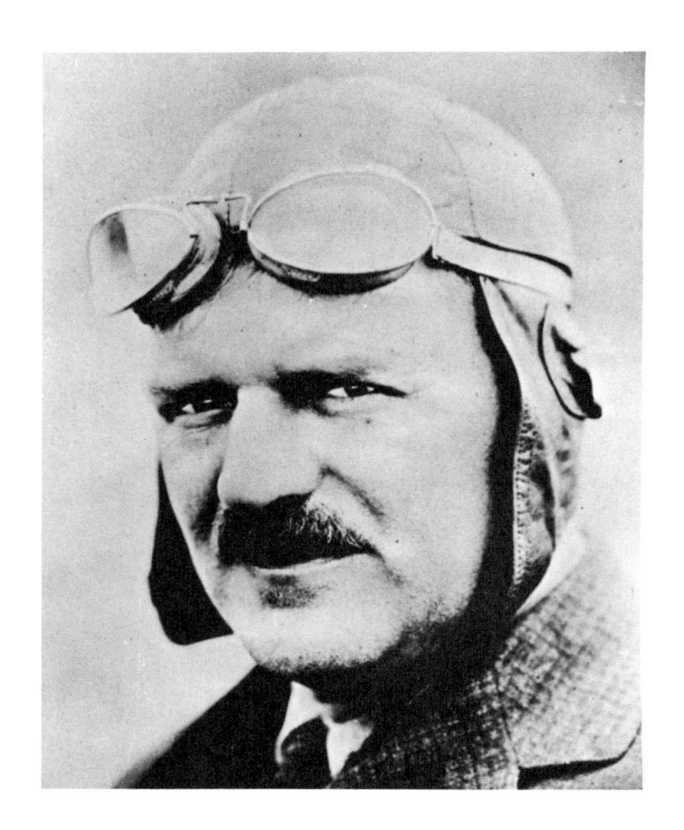

Louis Chevrolet in the early 1920's. IMS

Arthur Chevrolet in 1910. IMS

teenth Century governor of the French colonies in North America, which beckoned him. Soon engines became his consuming passion, and he learned many of their secrets working in the factories of Mors, Darracq, Hotchkiss, and de Dion Bouton. Then in 1900 he sought his fortune in the New World.

Louis spent about six months in French-speaking Montreal, where he worked as a chauffeur, an occupation which at that time included being capable of rebuilding an automobile from the ground up. Then he rejoined de Dion in its Brooklyn, N. Y., branch. This lasted until mid-1902, when he got a better job with the Fiat agency in New York City. He was the best mechanic they had, and it soon became apparent that he had exceptional talents as a driver of the unforgiving vehicles of the period.

On May 20, 1905, he was entrusted with a 90 HP Fiat race car and was entered in his first major contest at the old Hippodrome at Morris Park, N. Y. His opponents included many of the greatest drivers of the day, yet he won this important event. He was one of the first Europeans to drive in American competition and the burly, six-foot, 215-pound French-Swiss was a popular winner.

The Fiat was enlarged to give 110 BHP, and he was turned loose with it in the second Vanderbilt Cup Race on Long Island. He was a top contender until a front wheel collapsed, the juggernaut crashed, and he barely escaped with his life. But Louis continued racing the big Fiats in major contests in the eastern United States with consistent brilliance. He beat the formidable Barney Oldfield three times in 1905, which established him solidly as one of motor racing's true élite. His younger brothers, who had followed him across the Atlantic, continued to follow his example, became better-than-average racing drivers, and brought further luster to the Chevrolet name.

It was at the Morris Park race that Louis first met New Yorker Walter Christie, who both built and drove his own machines. His projects were fascinating, and in 1906 Louis left Fiat and joined Christie, whom he assisted in the building of a new front-drive car and in the preparation of a record machine powered by a 200 BHP Darracq V8 engine. Louis drove this car at Ormond Beach, Florida, to a new absolute record speed of 119 MPH, then turned the monster over to Darracq team driver Victor Hémèry. The clocks failed on his first run, his violently abusive French was coherent enough to the officials, and he was banned from the course. Thus his mechanic, Victor Demogeot, was clocked at 122.449 MPH and became the first man in history to travel at two miles a minute. That honor also could have been Louis' but it was not his duty to upstage a client.

Louis' reputation as a fast and fearless driver continued to grow, bolstered by his budding fame as a mechanic who had a special touch with high-performance machinery. This came to the attention of William Crapo Durant, whose booming young General Motors Corporation had absorbed Buick, a marque which then was actively building a reputation through racing. In 1907 Durant hired both Louis and Art, and by 1909 Louis' stature as one of the nation's foremost drivers was firmly established.

Durant lost control of General Motors for the first time in 1910, and without losing a beat he began forging yet another automotive empire. He needed new products with new names, and he commissioned Louis to create a Chevrolet car. By November of 1911 the prototype was running, and the Chevrolet Motor Company had been organized with the celebrated Louis Chevrolet as a substantial stockholder.

The original Chevrolet car went into production in 1912, and by the end of the year 2,999 of them had been sold—a very healthy start. But the big, six-cylinder phaeton carried an F.O.B. price of 2,150 dollars and Durant had made up his mind

In one of his first races, the Vanderbilt Cup of 1905, Louis crashed in his 110 HP Fiat and narrowly escaped death. MRS. SUZANNE CHEVROLET

that he wanted the Chevrolet to compete with the dirt-cheap Model T Ford. According to the legend, Louis insisted that his name be associated only with quality products, whereupon Durant reminded him that he no longer owned the use of his name. This led to a violent clash between the two men and to the abrupt end of their relationship. The legend has the virtue of simplicity but not of plausibility, and it is just possible that Louis, having served his purpose, was manipulated into taking the voluntary action which he took.

He broke with Durant very early in 1914, after which point he never had any further connections with the Chevrolet Motor Company, with one ironic exception which will be mentioned in due course. The experience so embittered him that he rid himself of all his Chevrolet stock (to Durant). If he had held onto it he would have become a millionaire many times over. Within six years Durant was nearing his first million-car Chevrolet year. Quietly buying up stock with his profits he was able

Louis in the 200 HP Darracq V8 at Ormond Beach in 1906. He clocked 119 MPH. MRS. SUZANNE CHEVROLET

to march into a General Motors board meeting in 1916 and announce flatly, "All right, gentlemen. *I* control this company." The company's profits for the three-year period just ended were 58 million dollars.

Louis had many friends, admirers, and powerful connections. One who was all three was French-born Albert Champion. When Louis split with Durant he determined to build cars on his own and to start with race cars. Needing a name, he founded the Frontenac Motor Corporation, and Champion financed him. Then, one day early in 1915, Louis found himself mortally wronged by this seeming best of friends. He crashed through the door of Champion's private office and beat the man—who was almost as tough as he was—almost to death. Louis told him never to cross his path again or he would finish the job. His instructions were observed by his firm's suddenly ex-backer.

So the Fronty race cars started in 1914 were not completed until 1916. They began winning races brilliantly, and in 1920 Louis was approached by tycoon Allan

Louis in the Christy front-drive, 1906.

A. Ryan and the directors of the Stutz Motorcar Company. As Durant had done and E. L. Cord was to do, Ryan saw rich potential profits in the biggest name in race-car engineering: at that moment, Frontenac. Ryan and his associates announced that they had raised a million dollars with which to back Louis in the production of a Frontenac passenger car: a fine, sporting, high-performance machine. All parties agreed that it should be an uncompromising thoroughbred.

Louis designed the prototype in collaboration with his close friend Cornelius Willett van Ranst, the ex-Duesenberg engineer who was in charge of engineering the Fronty four- and eight-cylinder Indianaoplis winners of 1920 and 1921. The car was highly original throughout. Its 197-cubic-inch, SOHC engine made the fullest possible use of aluminum, and the whole chassis was remarkable for light weight and efficiency. It was Louis' reply to Fred Duesenberg's Model A.

The whole Stutz technical and managerial staff observed the tests of the prototype at the Indianapolis Speedway in the late fall of 1921. When all the verdicts were in, Ryan clapped Louis on the back and laughed confidently, "It's a winner, old boy, and here we go!"

In the Darracq. MRS. SUZANNE CHEVROLET

Louis, ensconced in a fine, hundred-thousand-square-foot factory and lining up his staff of 1,500 workers, thought so too. After a long life of struggle, destiny had decided to allow him her smile. The optimism of the hour was summed up in advertisements which first appeared in January, 1922:

> Louis Chevrolet is offering the most remarkable motor of the season in his new Frontenac, being announced for the first time.
>
> The new Frontenac bursts into the automobile field with the momentum of its predecessors, who crossed the tape victorious world's champions in the past two International 500-mile races at the Indianapolis Motor Speedway. The Frontenac is the latest creation of Louis Chevrolet, master designer-builder-driver.
>
> The only man who ever built two different cars in two consecutive years to win the 1920 and 1921 International 500-mile Sweepstakes races, has given more than 20 years of his life, experimenting and preparing for the car he is now offering to the public.
>
> It is undoubtedly the cleanest motor ever displayed: It has speed and power: Its vitals are unusually accessible. It is the most conveniently arranged power plant you ever have seen.

But destiny's favor had not been all that sincere, and another trap had been set for Louis, with his usual proneness for this sort of thing. When he made his contract with Ryan he did not bother to retain the services of a lawyer. Also, he held onto the Frontenac Motor Corporation of Michigan, while Ryan went ahead and organized the Frontenac Motor Corporation of Delaware. Then, just when things looked brightest, a major economic depression erupted, absolute hell broke loose on the Stutz-Ryan-Wall Street axis, and the Delaware corporation went bankrupt. Since Louis' orders to suppliers did not specify *which* Frontenac Corporation was committing itself, he was left holding the bag for all Frontenac indebtedness. Thus the Frontenac sports car project died aborning, and on April 16, 1923, Louis filed a voluntary petition in bankruptcy in the Indianapolis Federal Court. The assets of the Michigan corporation were 425 dollars, its liabilities 88,163 dollars.

Van Ranst kept Louis' fortunes afloat with the creation of the Frontenac cylinder head for the Model T Ford engine, and these heads sold by the thousands for years. With brother Art in charge of the Fronty head program, Louis again devoted his energies to bigger things. This time he tried to attack the passenger-car field by offering it an engine that was remarkable for efficiency *and* silence. He obtained the American manufacturing rights for the Argyll single-sleeve-valve engine. He built a complete car around this power plant again with the intention of producing and marketing it under the Frontenac name. But he was unable to complete the financing of this project before his Argyll option expired and was snapped up by Continental Motors.

However, the aircraft industry was beginning to flourish and there was a need for powerful and efficient engines for light planes. This was a ready-made opportunity

The first Chevrolet. Cliff Durant at wheel, Suzanne Chevrolet beside him, Louis in white coat, 1911. MRS. SUZANNE CHEVROLET

for a man with such a distinguished record in the high-output engine field, and in 1926 Louis picked up this gauntlet. He built prototypes of very compact DOHC, air-cooled, inverted four- and six-cylinder aero engines which passed all civilian and government tests. Who provided the engineering talent on this project is not clear, but soon both Louis and Arthur were claiming—independently of each other—to have designed the "Chevrolair" engines. In 1927 the two brothers scrapped their lifelong partnership. Art retained the speed equipment business but renamed it the Arthur Chevrolet Aviation Motors Corporation with production of the Chevrolair in mind.

Louis found an ally in an ex-Ford dealer named Glenn L. Martin, who also had great visions for the future of aircraft. Louis joined him in Baltimore where, reminiscent of his old deals with Durant and Ryan, he formed the Chevrolet Aircraft Corporation with Martin. Right on cue the Great Depression arrived, wiping out all prospects for the success of the company in the visible future. Louis ceded his interests to Martin, who formed the Glenn L. Martin Motors Company. The four- and six-cylinder engines were produced as the Martin 4-333 and 6-500. Thus Louis seems to have contributed to the formation of yet another empire. But today the Martin Company is officially unaware "of business links between the aircraft pioneer Glenn L. Martin and Louis Chevrolet."

Then there was Dominion Motors of Toronto, the successor to Durant of Canada. In 1931 a Dominion press release announced to the world:

Van Ranst and Louis with their SOHC aluminum engine for the Stutz-backed Frontenac passenger car. VAN RANST

One hundred fifty Frontenac Six automobiles left the plant of Dominion Motors on September 3, initiating a new name in the automotive industry.

This was the American Continental-engined De Vaux car which was marketed in Canada under the Frontenac name which Louis had given life and greatness to more than a decade before. This was hardly a model of ethical practice but Louis was helpless; his legal rights to the name were valid only in the United States. Several thousands of these Frontenacs were sold before Dominion went bankrupt in December, 1933.

Legend has it that in 1933, when jobs were terribly difficult to get, Louis managed to get one as a common mechanic in one of the Chevrolet factories. Perhaps his name helped.

The following year really crushed the fifty-six-year-old warrior. Then the older of Louis' two sons died, and this broke his heart. After that a sister's house burned to the ground. It had been a roomy home in New Jersey where Louis had stored all the drawings, documents, and memorabilia of his tumultuous life. Louis capped these events with a cerebral hemorrhage after which he was subject to paralytic attacks for the remainder of his life.

Louis and his beautiful wife, Suzanne *née* Treyvoux, a French girl whom he had courted and wed during his Fiat campaigns of 1905, retired to a small apartment in Florida, where the climate put a minimum of strain on his fragile health. Then, in 1941, fate brought the battered little couple back to Detroit. There Louis underwent a leg amputation, from which he did not recover. He left a very unkind world on June 6, 1941, and is buried beside his brother Gaston in Indianapolis.

Van Ranst, who was to make many important engineering contributions to the nation's automotive and aircraft industries, spent four years working for Louis. They became like father and son, and their profound friendship was lifelong. Van Ranst said:

> Louis was the leader of the family. Arthur and Gaston were fair mechanics but I never considered either of them great drivers. Louis was good at everything and had all the talent in the family.
>
> I don't believe one could classify him as an inventor or designer, but over the years of contact with many makes of automobile Louis had a native ability to separate the good points from the bad and with a memory like an elephant was able to retain his observations and use them to good advantage. He dictated to others *what* he wanted, not how it was to be done.
>
> Louis had terrific drive, along with a highly competitive spirit and the ability to get things done in the shortest time. He was not a finished mechanic, often sacrificing quality for time. He had a violent temper when rubbed the wrong way, but under normal conditions was as mild and gentle as a lamb. He was one of the greatest men I have ever known and I believe I owe a large part of what little success I have had to his tutoring.

chapter 7

HARRY ARMENIUS MILLER

HARRY MILLER WAS born in Menomonie, Wisconsin, in 1875. His mother was Canadian. His father, who spelled the name Mueller, was born in Germany, where he trained for the priesthood and, on the side, became an accomplished musician, painter, and linguist. In Menomonie Mueller worked as a schoolteacher to support his brood of two girls and three boys. His son Harry deserted school for a job in the town's one machine shop at the age of thirteen.

Mueller was furious. He had high hopes for the careers that his sons would pursue in this land of opportunity, which Harry seemed determined to throw away. But the only higher learning that had any meaning for Harry had to deal with machinery. He quickly learned the machinist trade and then learned to maintain the steam donkey engines in the town brickyard and in nearby lumber camps. He was in his element. At seventeen he was ready to tackle wider horizons and left home.

He drifted to Salt Lake City, then home again, and then in 1895, when he was nineteen, he decided to give Los Angeles a try. There he got a job in a bicycle shop and met Edna Lewis, who was all of sixteen. They fell madly in love and became engaged.

While waiting for Edna to reach the age of consent Harry worked at his 18-

Harry Miller and the first 91 cubic-inch engine. TED WILSON

dollar-a-week job and on the side set up a tiny shop where he fabricated special parts for converting ordinary bicycles into racers. He saved his money, and when they were married in 1897 Harry took his bride back to Menomonie for an extended stay with his family.

Now he was a sophisticated young man from the big city, and he had his pick of jobs for which his experience qualified him. To get to and from work he designed and built a bicycle on which he mounted a small one-cylinder engine. It has often been claimed that this was the first motorcycle in the United States. If so, it did not occur to Harry to protect the idea, and soon there were companies making fortunes from it. Edna said:

> Harry was always working. Always building things or thinking about things to build. That's all he ever did or cared about.

Then, according to a 1937 newspaper biography:

> The following summer at Menomonie he worked out a peculiar four-cylinder engine, clamped it on a rowboat and showed his cronies how to enjoy their afternoons off. It was the first gasoline outboard motor in the country.
>
> "About that time my wife got terribly homesick and we packed up and went back to California," Miller recalls. "I never did get around to patenting the outboard engine—clean forgot all about it."
>
> A young machinist in the shop at Menomonie was not so lackadaisical. He brought out a two-cylinder outboard shortly afterward. His name was Olie Evinrude, and he became the papa of a highly profitable industry.

It was in 1900 that Harry and Edna arrived in San Francisco, where he got

This 1916 photo of the Miller carburetor factory indicates that the name was already intimately associated with racing. TED WILSON

another machine-shop job and built another motorcycle to get around on. Here he got experience with foundry work and with the manufacture of pistons. Later, shortly before World War I, he set up his own piston factory, and it seems to have been the first in the United States to make pistons of aluminum. From San Francisco Miller moved back to Los Angeles, where he designed a novel type of spark plug. It must have been protected because he sold the manufacturing rights to the Peerless Motor Car Company for a good price.

In 1905 Harry built his first automobile. All that Edna could remember in later years about its structure was that it had a dog clutch and no transmission. He used to treat Edna and their son to Sunday drives, and getting started or stopped was usually an adventure. Edna marveled at his inventiveness and told him that he could be a great captain of industry if he would only aim for that goal. The thought repelled him completely.

Just what he wanted out of life he never discussed with anyone, including his wife. He liked money and wanted to have lots of it for good but unostentatious living. When Edna would panic about finances he would just shrug and say, "Oh, I'll make some more."

Which he always did, for a long long time.

Edna realized that her husband was "a funny man," meaning a strange one. He required almost no sleep and would often work for three days and nights straight through. At home, he would sit and think or lie awake in bed and think for hours at a time, as though in another world. This was in the most dramatic contrast with his behavior when the meditative mood was not upon him. Then he was all bright spirit, affection, and good-natured mischief.

Some of his pranks were not at all funny to Edna. He would tease her by taking whole series of statements out of her mouth before she had uttered them, and this frightened her. Other aspects of his seeming clairvoyance were more disturbing. He had a knack for sensing the date and hour when people would die. Walking down a Los Angeles street one day they passed a department store which had a portrait of its founder done in mosaic in the sidewalk at the main entrance. Harry paused, studied it, and, moving on, said "That poor devil will be around for just two more days."

When the newspapers appeared, carrying the notices of this total stranger's death, Edna laid down the law, and Harry closed that side of his nature to her.

Harry and Fred Offenhauser had tremendous rapport but on a very nuts-and-bolts level, and the Old Man never mentioned the occult to Fred. But he did to Leo Goossen. One day Harry, clutching an old envelope covered with his da Vincian scrawls, said, "Leo, listen to me and try to understand this. *I* don't do these things. I get help. Somebody is telling me what to do. I mean it, Leo. I rely on it."

Whatever his reaction may have been, Goossen was not disturbed. Who knows where *any* ideas come from, he seemed to feel. When practical engineer Eddie Offutt came to Miller from Louis Chevrolet in 1923 a profound bond grew between the two men. Again, Miller admitted having his "control," and Offutt accepted the statement without showing any interest in trying to judge that about which he knew nothing. Harry never mentioned his "control" to his wife.

In his heyday Miller stood five feet six inches tall, had black hair, blue eyes, a fair complexion, ruddy cheeks, medium build, and a slight paunch. He favored gray suits, black shoes, boutonnières, a small mustache, and expensive hats, which he wore with verve. He presented an impression of easy affluence, lively good taste, and conviviality mixed with an almost boyish shyness.

Of his taste, said Ed Winfield, who got his start with Miller:

> He was the originator, in the automotive field, of doing an artistic job on his machinery. He was more of an artist than an engineer. The one thing he insisted on—and that was pure Miller influence—was having everything well-proportioned and well-finished, regardless of structural quality. Others around him, like Fred and Leo, could worry about that.

Harry's shyness was matched by equally childlike frankness and candor. Said Edna:

> If he didn't like somebody he'd tell them so to their face. It could be God or the President; it didn't matter. He once said to one of the richest men in America, a man who wanted to build a team of cars to race in Europe, "Why, man, you talk engines like a fool!"

He liked and often felt a deep fondness for most of the men who worked for him because they were the best that he or Fred could find. They were excellent at their respective callings, the callings that made sense to Harry. They often would become outraged by the Old Man's cavalier style—such as blowing all of the payroll money on his latest mechanical whim. They would quit, vowing never to speak to the man again. But he was the enchanted king of that esoteric world, and most of them knew it and came back in spite of themselves, being unable to bear not being participants in that fantastic experience and dream. When Harry saw them back in the shop he always acted as though they had never left.

He was utterly indifferent to adversity. One day Offutt, heartbroken, came to the Old Man and informed him that he had just blown up 5,000 dollars worth of engine on the dynamometer. Miller asked how bad it was and Offutt told him it was totalled. Miller shrugged philosophically and said, without a trace of rancor, "Well, let's see how fast we can build another one."

He would never criticize a man for making a part too light, but woe to the man who made any part heavier than it had to be. Any man who failed to get this message did not last in the organization.

Miller's shyness was due in part to self-consciousness about his vast lack of formal education. It did not hinder his flow of ideas, but it did inhibit his ability to give them expression. In trying to solve technical problems he would constantly collide with, for example, his ignorance of mathematics. On such occasions he would call for the help of a trusted associate, like Goossen or Offutt, who could give him his answer with a flick of the slide rule. But there is no record of anyone ever boasting

Barney Oldfield (left) and Miller (right) with model for an unrealized land-speed record project in late 1920's. TED WILSON

that he *helped* Harry Miller. Perhaps because he was an awesome genius, the men close to him were grateful for his human limitations, which made it possible for them to share in his creative flights.

He was not remotely a businessman. He could not run a business but would let no one run his business for him. He trusted no one, and even his wife never had a hint of what he earned or owned other than that he kept making and losing six-figure fortunes. She had a normal instinct for economic security and would beg him for more than her household allowance so that she could start a savings account. He would just say, "Honey, I make it and I'm going to spend it. As long as you have everything you want and enough to run the house, that's all right."

In the mid-Twenties Harry bought a wonderful ranch in Malibu Canyon in the Santa Monica Mountains, about fifty miles from the city. Both he and Offenhauser loved the outdoors, and they used to hike back into the hills and then sit and talk for hours. Fred knew his boss all too well and worried about him. On one of these occasions he told Harry flatly that he would be more than wise to set up a trust fund for himself and to put all his real property in Edna's name. "And have her throw me out?" Harry roared. It was three weeks before he would speak to Offenhauser again. Fred knew, as Harry was somehow unable to know, that Edna loved her husband a great deal more than she loved herself.

An index of just how golden the golden era was, was Miller's prosperity during the Twenties, derived exclusively from the manufacture and sale of racing cars and engines. He priced his products in neat, round figures: 5,000 dollars for a supercharged 91 engine, 10,000 dollars for a complete 91 rear-drive car, and 15,000 dollars for a front drive. His gross earnings for the 91-cubic-inch period alone were close to a millon very sound dollars, perhaps more. Then in late 1928, when the doom of thoroughbred racing seemed sealed, Harry went to work on creating a future for himself that would be as meaningful as his past.

Like Louis Chevrolet, Harry saw the signs of impending revolutionary change and opportunity in the aircraft industry. What he foresaw was vast and completely unlike the cottage industries he had always known. So, with financier George L. Schofield, Fred E. Keeler of Lockheed Aircraft, G. E. Moreland of Moreland Aircraft, and Gilbert Beesemeyer of Bach Aircraft and large petroleum interests, he formed the Miller-Schofield Company, capitalized at 5 million dollars. Miller received cash consideration of a reputed 100,000 dollars along with a substantial interest in the new organization and its designed-in success. Its immediate objectives were the production of ultra-efficient aircraft engines and the mass production and nationwide distribution of Miller carburetors, light alloy pistons, and pushrod OHV cylinder heads for the Model A Ford. A nine-acre site was acquired, and work commenced on the factory in which these diversified products would be made. Meanwhile the old Miller plant operated at full tilt trying to keep up with ever-growing demand by turning out fifty Model A heads daily for several weeks (these Miller-Schofield heads later became famous under the name Cragar). Then the Great Depression struck; the whole dream dissolved into nothing, and in mid-1930 Miller-Schofield was bankrupt.

Miller had set up his own small machine shop, which he called the Rellimah Company, inverting the name and initials that were part of the Schofield deal. When it failed key men who also had been part of the deal, such as Offenhauser and Leo Goossen, rejoined the Old Man. But the prosperous days were gone, and Miller's fortune quickly trickled away. In 1932 he, too, found a sheriff's padlock on his door.

Harry took his bankruptcy and the ruin of his life's work very hard. He went East in a frustrating search for new fields to conquer and, although he had become afflicted with diabetes, he took to drinking heavily. At Indianapolis in 1934 he saw Walt Sobraske, who had worked for him as a schoolboy and to whom Miller had truly been like a father. Walt was a member of the little group of talent that Offenhauser was struggling to hold together. Harry rushed up to him and embraced him. Then, holding him at arm's length, Harry looked steadily at him and said, "Son, one of these days I'm going to come back and pay every one of you boys every cent I owe you."

Then his eyes flooded with tears, and he wheeled and lost himself in the crowd. Walt never saw him again.

Boat-racing champion Dick Loynes recalled that there were times when you could send a parts order to Miller and be lucky to have it filled in two or three months. It didn't interest Harry that there was good, ready money to be made in the replacement-parts end of his operation. Loynes said, "He worked at what he wanted to work at, limited only by the need to eat as he went along."

But at the same time and to Offenhauser's despair Harry would cheerfully ship parts to any deadbeat who put on a good show in racing. He knew that he never would be paid and all too many people knew that he really didn't care, and they traded on this. When Fred would protest such a shipment Harry's typical response would be, "What? And not have our cars out there running?"

Harry was passionately intolerant of any meddling with his creations. Valve stems and connecting rods in the 91-cubic-inch straight eight were weak points of the design which Miller was content to accept. When he learned that Frank Lockhart, who was campaigning one of the factory-owned cars in the East, was making his own valves and rods he snorted, "Why does every kid who wins a race suddenly think he's an engineer?"

He wired Lockhart, forbidding "the kid" to make any modifications to the machine. Lockhart responded by simply buying the car, and then a second one as a spare. With "the kid's" modifications they became the most phenomenally successful cars in American racing history.

Miller was richly endowed with negative talents. He was a hopeless procrastinator and had the reputation of never having been on time in his life. He remained interested in a project only until he conceived a new one, which was far too often for his own good. He would walk out on the most complex undertakings, saying to some underling, "This is what I have in mind. Take over and finish it."

This walking-out was one of the themes of his life story, and as a result countless wonderful ideas never reached fruition. This is why the straight-eight 91 was his greatest engine. He just happened to stick with it and to make enough copies for it to become properly debugged, refined, and perfected. There is a vast catalog of magnificent Miller failures that never had a whisper of that chance.

One of Harry's great loves was animals. From the early days in Los Angeles he kept a parrot in the drafting room. It flew freely there, would attack Offenhauser's bald head whenever he would enter, and contributed an air of confusion that delighted the Old Man.

He loved monkeys and always would have at least one around him, riding with him in his car, clowning in his office, and raising absolute hell in Fred's orderly shop.

His Malibu ranch was a veritable zoo. There was a mob of dogs, each named for some celebrated friend; Rick was one, named for Eddie Rickenbacker. There were gangs of monkeys, plus foxes, coyotes, opossums, deer, and a black bear—a regular Noah's Ark. He had a wire-mesh birdhouse that was about forty feet square, twenty-five feet high, and enclosed a good-sized oak tree. There he kept every bird he could buy. Chickens and ducks waddled around the yard, and he had a license to raise quail, which he did.

It was a nice place in a green glen surrounded by wild, majestic mountains. The house had three bedrooms, and its two wings framed a large patio with an outdoor fireplace and barbecue pit. Beyond, there was a pretty fruit orchard, where Harry enjoyed driving his tractor. Here it was his delight to share his good luck with friends. A caretaker kept the place shipshape at all times, and nearly every weekend there

Miller with Paul Whiteman mustache and one of the last of the 91 engines. The camshaft blower drive, blower, and intercooler were experimental and are atypical of Miller practice. GB COLLECTION

was a party at the ranch for the racing fraternity. There were seldom fewer than thirty guests, and Harry and Edna wined and dined them royally. This is what money was for—this and the machine.

After his bankruptcy Harry's luck never improved. Asked if he thought Miller had lost his "control" Offenhauser snorted, "Control, hell! He just lost his team. That's all."

And everyone, but *everyone,* agrees with this.

Without the old team he designed and built some glorious cars and engines but they rarely ran well, and they were plagued by ghastly luck. Somewhere between the Ouija board and the drawing board, between the team and the lonely Old Man, there were some vital elements missing. Perhaps the most important lack was that of the trusted friends he could turn to for help in giving his ideas expression. They would tell him when he was wrong. In the East he found only yes-men.

He drifted from one project to another and finally landed in Detroit. The Motor City had never had any fondness for Harry Miller (incredibly, there is scarcely a reference to him in the *SAE Journal*) because he was a source of embarass-ment to the industry. When its best engines were offering the public all of 0.25 HP per cubic inch Harry's purebreds were pushing, in some cases, 2.75. He put much of the contempt into the words, "Detroit Iron." He found no haven, no proper employ-ment there even when the city was glutted with cost-plus war contracts, and every machine-shop owner was getting rich. He had to take in petty job work to keep his tiny shop going. And there were other reasons for his not being wanted. He was old. He was a failure. And there was his disfigurement.

It started as a thorn-prick when he was hiking behind the old ranch. It slowly turned into a cancerous growth on his face. It was a grim thing, and he demanded that Edna remain in Los Angeles rather than share his—to him—revolting condition. She fought this edict with all her might, but he would have it no other way. He finally had the growth removed by surgery, but he remained in constant pain and never left his room without a dressing on his face. He grew weaker and his heart began to fail. For companionship he had the usual clowning monkey. And for a friend he had Eddie Offutt.

Offutt's own engineering career had brought him to Detroit, where he put in long hours on a military project. He promptly found Miller, who had been like a father to him, too. Not just gladly, but with a sense of profound privilege, he found eight hours in every day to donate to the Old Man's latest project.

Offutt described it to me in 1952—Harry's last drawing-board gamble, a design for a revolutionary type of small, inexpensive production car. Today we can say that it was a thorough anticipation of the Issigonis BMC Mini, but more advanced in many ways. Harry had designed an automatic transmission for it, and this and many other components already had been built and tested in prototype form. Then Offutt had to wire Edna that her man was dying. His struggles ended on May 3, 1943, before Edna could reach his side. His friend to the last said, with quiet, profound feeling:

> That man could see so much farther into the future than any of us. He had the courage and capacity to dream in the grandest manner. That was his life's crusade and that is why we worked for him and felt toward him as we did.

chapter 8

LEO WILLIAM GOOSSEN

IT IS NOT INEVITABLE that the dedication of one's life to the machine must result in heartbreak and tragedy. For many men it is the profoundest possible fulfillment of vital purpose and, even if the material rewards turn out to be meager, the satisfaction in life is beyond price. Such has been the career of Leo Goossen—a career which has provided an amazing measure of the substance of American motor racing history.

Engineer John R. Bond wrote in *Road & Track* for April, 1964:

> To me the man who makes the layout drawings is the only real designer . . . . Prior to the Ford twin-cam project there was *one man* in all the United States who earned his living as a racing engine designer. He is, as some of our readers know, Leo Goossen.

Leo has been America's one "real" designer of racing machinery, in John Bond's sense, for many more years than he cares to remember. It is not vanity that makes him want to avoid discussing his age; it is society's attitude toward ripe years that makes him want to forget the subject. Society says you should be put out to pasture at a certain point. But the history of art, science, and technology is crowded with examples of great talents that defy the common rule and only continue to flower with the march of years. Leo's is such a talent. If it has been denied the recognition it deserves and if it has been frustrated in its fullest expression, its owner nevertheless can look back on a career of really magnificent achievement. And, if Leo has his way, that career still has a long way to go.

Alden Sampson and Leo Goossen with V16 engine based on Miller 91 components. Note Leo's sensitive hands. TOM STIMSON

Both Leo's parents came from Holland and settled in Kalamazoo, Michigan, where he was born. Then they moved to Flint, where the lean, gangly kid got halfway through high school before having to go to work. His ambition was to become a draftsman, and his first job, in Buick's blueprint room, seemed like a good start. He studied hard on his own, took night classes, and paid a tutor to teach him mathematics and engineering fundamentals. In six months he was promoted to the status of record clerk, taking care of the drawing files.

Buick was no giant then, and Leo moved freely among the twenty or so draftsmen in the engineering department. Its head was E. A. de Waters, soon to become the firm's chief engineer. One day Leo, then sixteen, worked up the courage to ask if his boss would mind looking at a few of his drawings. De Waters admired the promise he saw in the boy's work, took him under his wing, and brought him to the attention of chief engineer Walter L. Marr, designer of the original Buick engine. Leo's ability, plus his gentle, earnest, and dedicated spirit, attracted Marr, who made him his protégé. Leo's first taste of racing design came when he was assigned to make tracings of the engineering drawings for the Buick *Bug* transmission. Then Marr turned him loose on the layout and detailing of an experimental four-cylinder side-valve engine; and he had the thrill of seeing two prototypes made and run, and of often rubbing elbows with the aces of Buick's racing team, the great Louis Chevrolet and Bob Burman.

In 1917 Marr's health took a downward turn, and he moved headquarters to his estate at Signal Mountain near Chattanooga, Tennessee. Marr took his favorite draftsman with him, and there Leo laid out another experiment, this time a side-valve V12. Again, two prototypes were built, and today one of the resulting V12 Buicks is still in the possession of the Marr family. And so, incidentally, is Leo's original drawing board still in the room which he occupied there. The following year his major project was translating the V12 into a V6 configuration. One prototype was built and installed in a car, which may still exist. Leo commuted to Flint, carrying drawings to Buicks general manager, Walter P. Chrysler, who called him "son." He still has Chrysler's strong letter of recommendation given to him when he had to leave his employer under painful circumstances.

It was at the very start of 1919, in the dead of winter, that Leo came down with a severe respiratory ailment. His doctor found a spot on one of his lungs and, tuberculosis being the terror that it was then, urged Leo to get himself to a warm, dry climate if he valued his life. He landed in Silver City, New Mexico, where he rode the range for half a year and where his spot and symptoms disappeared. Liking the Southwest, he pushed on to Los Angeles to find a job.

There weren't too many opportunities in the Far West in those days for an experienced automotive draftsman. But there was Harry Miller, who, with his shop foreman and production manager, Fred Offenhauser, manufactured carburetors on a big scale, repaired racing cars on the side, and had taken a few flings at designing his own engines and chassis. Leo wandered into Miller's office in August of 1919, produced his letter from the great Chrysler, and was hired on the spot for 30 dollars a week.

There was a job to go right to work on—the radical TNT sports-racing prototype (see Chapter 11). Miller had hired three other draftsmen to rush this project to completion in time for the 1920 Indianapolis race, but industry-trained Leo was put in charge of the team. In such critical areas as the design of bearings, gears, and cams he was in possession of the most advanced ideas of the day, and Miller was delighted with his windfall.

Then the real thoroughbred period began when Tommy Milton and Ira Vail came to Miller, wanting an engine that could beat the Duesenberg straight eight. Leo went to work on the drawings early in 1920, and in 1921 Vail finished seventh at Indianapolis with one of the first Miller 183's. According to Milton, Leo's talents were decisive in finally unlocking that engine's vast potential.

Then, still in 1921, Cliff Durant commissioned Miller to build a small fleet of six complete two-man, 183-cubic-inch race cars for his "Golden Gang." This was Leo's first experience in drawing and detailing a complete racing vehicle, right down to its axles, brakes, steering gear, frame, tanks, and body. Leo was in his heavenly element, and his drawings were lyrical. Those cars were engineered to the last nut, bolt, and bracket, and each piece was admirable for its economy of material, grace of line, and saving of weight. Then Jimmy Murphy's 1922 victory at Indianapolis with a

Miller 183 engine in his French Grand Prix Duesenberg chassis brought a flood of orders to the little Los Angeles factory.

In July of 1922 Leo began drawing up a 122-cubic-inch engine for the 1923-1925 formula. Miller decided to switch from three to five main bearings and from four valves in a pent-roof combustion chamber to two valves in a truly hemispherical one. With these points resolved all the rest was Leo's responsibility: clearances, dimensions, shape of housings, gear-tooth forms, everything. As engineering genius Vittorio Jano once told me:

> Coming up with ideas doesn't have to be a remarkable process. It's easy to suggest mining metals on the moon. The men who work out the techniques for the realization of the raw idea are the real achievers.

By this time Miller, who was more than shy about his lack of technical background, had given up any pretense of checking Leo's drawings and calculations. Leo had demonstrated that he was a master of his art and that was all that the Old Man needed to know. During that same period Leo laid out a completely new chassis for the Miller 122, again engineered to the finest detail. It was with one of these cars that Milton won the 1923 Indianapolis race and spurred the stampede toward these unique machines that required no emblem to mark their identity.

Early in 1924 Jimmy Murphy's skillful mechanic, Riley Brett, convinced Murphy that there were great possibilities in the front-drive principle. Murphy assigned Miller the job of creating a front-drive race car, and Miller had the naïveté to propose and launch work on a transverse-engine approach to the problem. Being a prophet too long after Christie and too far ahead of Issigonis, his approach was spurned and scorned. Somehow, between Miller and Goossen, a brilliant alternative solution was devised. The resulting revolutionary and pioneering Miller front drive had immense repercussions upon the world automotive industry and, among other things, marked the world's first use of de Dion suspension in a race car. It is ironic that Miller, who hardly ever bothered to patent his ideas, thought enough of this one to cover it thoroughly, to his considerable personal benefit. Leo was unaware of this patent's existence until I showed him the documents thirty-five years later.

After Duesenberg's triumphant introduction of supercharging at the Speedway in 1924, Leo and Miller quickly became masters of this art. During that year Leo also laid out an eight-cylinder opposed DOHC aircraft engine for Miller. Just one prototype was built.

After the 1925 Indianapolis race Leo began drawing the 91-cubic-inch Miller straight eight. This was a scaled-down and tremendously refined version of the earlier 122. Miller of course ordered the job, but the perfection of its execution—esthetic as well as technical—was due largely to the sensitivity of the hand and the mind which drew it. In both minds, technical excellence was incomplete without visual excellence. And even Offenhauser, who fought stubbornly against impossible odds to make his shop run at a profit, agreed that the summits of achievement in this most esoteric of fields could be gained only through adherence to ultimate standards of quality.

In 1926 Leo drew up the Miller 310-cubic-inch straight-eight marine engine, the first one for famous boat-builder John Hacker of Detroit. It and the Miller 91 were highly successful both in national and international speedboat racing. Then Leo designed a four-cylinder, 151-cubic-inch version of the 310. This was Miller's most profitable marine unit, and about seventy-five were sold at 1,500 dollars per copy. Then Leo adapted many of the 151 and 310 components to a 620-cubic-inch V16 for racing boats of the Gold Cup class.

In 1927 Leo drew up the numerous special features of the front-drive 91 with which Leon Duray set many national and international records, including the world's closed-course record of 148 MPH which was set in 1928. He also laid out and detailed a pushrod straight eight for Durant Motors in 1927. Only one prototype was built.

In 1928 Leo laid out an opposed eight-cylinder marine engine for the Richfield Oil Company's racing division. Then, a huge W-shaped 24-cylinder, 5 by 7 inch bore-and-stroke marine racing engine based on Liberty aero engine components. Before this he had already resolved the basic design problems of the front-drive L-29 Cord car.

The Cord project was one great complex of compromises, all related to manufacturing cost. After it came the cost-no-object commission to design and build the Phil Chancellor front-drive sports car. Its power plant was a V8—one half of the 620-cubic-inch marine V16. For the first time in Miller front-drive practice, transmission was placed forward of the front axle shafts. Leo designed this chassis from scratch, and it was a milestone in front-drive evolution.

When Harry A. Miller Incorporated merged with Schofield in 1928 Leo was one of the assets in the bargain. For the new firm he designed what was to become known as the Cragar head for the Model A Ford engine; for it he utilized stock Buick rocker arms. It made a bomb of the A-bone. Also for Schofield, Leo designed a gear-driven DOHC head for the Model A. Only about three were made, but they made racing history at Ascot and on other Southern California tracks. He did this in 1929, along with designing the Hickey piston-valve straight eight.

After Schofield's bankruptcy in 1930 Leo rejoined Harry Miller. His first job with the Rellimah Company was to draw up a 230-cubic-inch straight eight for Louie Meyer. It had an aluminum cylinder block, detachable head, and vertical "downdraft" inlet ports. Several other engines drawn by Leo around this period used this revolutionary porting arrangement. He used it first in the V16 engine which he drew for Riley Brett in 1929. It was a brilliant idea, and Leo complimented Brett for having conceived it, but Brett honestly admitted to having seen it on some nameless one-off engine in the East. The "downdraft port" has become a feature of many of the world's best high-performance engines, the most recent of which is the DOHC Ford.

For Leo there was no transition between the various epochs of American racing; he just continued practicing his art, which was the design of machinery for ultimate performance. In late 1930 and early 1931 he laid out the 200-cubic-inch Miller four to

Leo Goossen in 1965; alert and full of fun and kindness. GB COLLECTION

The Miller drafting room about 1924, where Miller's parrot had the freedom of the air. Leo is second from the rear. GB COLLECTION

compete with the DOHC Ford conversions that were tearing up dirt tracks at that time. This engine became the Offy 220. During this same period Leo laid out the Miller rear-drive de Dion chassis.

The next year was a very busy one. In March Leo began layouts for a 1,113-cubic-inch marine racing engine for Gar Wood. These huge V16's had a Roots blower for each bank of cylinders. Later in the year Leo enlarged the Meyer-type 230-cubic-inch straight eight to 249 and then to 268 cubic inches, substituting cast-iron blocks for the original linered aluminum. Then he drew up the four-cam, 308-cubic-inch V8. Then, extending into 1932, he detailed the remarkable Miller four-wheel-drive chassis, which utilized the V8 engine.

In early 1932 Leo achieved miracles by co-designing with Harry Hartz and laying out the perennially successful Hartz-Miller 183 straight eight; Miller's only connection with the project was his blessing. After Miller's bankruptcy Leo spent the summer with him in Detroit, where the Old Man and his friend Preston Tucker tried vainly to promote new activities. Then Leo returned to Fred Offenhauser, who had bought some of the old Miller drawings, patterns, and machine tools and struggled to keep the tradition alive. For the new, little Offenhauser Engineering Company he scaled the 220 four-banger up to 255 cubic inches and then drew up the Dankwort two-cycle Diesel opposed engine. Then in late 1934 and early 1935 he designed the Pirrung front-drive race car, including its novel transmission. Also in 1934 he laid out and detailed the immortal Offy Midget engine.

In the summer of 1935 he drew up a SOHC midget engine for Riley Brett and then set up his own design office, where his first projects were a single-cylinder two-stroke and a V-twin motorcycle engine for Ruckstell-Burkhart. Leo's redesign of the clutch of Captain George Eyston's land-speed record machine was decisive in enabling new absolute records to be set at Bonneville that year, as Eyston himself announced in a BBC interview.

In the summer of 1936 Leo started layouts on a two-valve, hemispherical-chamber DOHC four for Offenhauser. Then, and until May of 1937, he worked on the layout of the Thorne Six engine and chassis for Art Sparks. It broke the Indianapolis track record that year. For Dutch Drake, Dale Drake's brother, Leo drew up an improved version of the British single-cylinder JAP (J. A. Prestwich) engine for short-track motorcycle racing. Then, between summer 1937 and May 1938, he laid out and detailed the engine and the entire chassis, including four-speed transmission, for the 183-cubic-inch Bowes Seal Fast straight eight which was campaigned by Louis Meyer and Rex Mays.

In the summer of 1938 Leo laid out a 255-cubic-inch hemispherical-chamber six in the Miller idiom for Joe Lencki of Chicago; with 400 smooth horsepower it still was in contention a quarter of a century later. In the fall and for Riley Brett he laid out the entire Sampson chassis around the ex-Frank Lockhart sixteen-cylinder engine. In 1939, again working for Brett, he co-designed the Sampson midget engine, a considerable improvement over the Offy.

In the spring of 1940 Leo drew up a single-cylinder piston-valve Diesel engine for Hickey. In the fall he co-designed the Novi supercharged V8 engine. All drawings, layouts, and calculations were made by Leo in his home and on the same drawing board on which he had laid out the original Miller 183. Watching it all was the old shop parrot. Miller gave her to Leo in 1932, and he still has her today.

In 1942 he laid out the transmission for Lou Moore's immensely successful Offy-powered front-drive Blue Crown Specials. Then, and into 1943, he did another job for Hickey: a four-cylinder opposed two-cycle Diesel with variable combustion chambers. In 1944 and 1945 he worked on the Ruckstell two-cylinder four-cycle APN engine for the United States Army. Then, in 1945 and 1946, he co-designed, laid out, and detailed the Novi front-drive race-car chassis, which was decidedly original throughout. In its first appearance at Indianapolis it broke the track record.

In the spring of 1946 Offenhauser sold his business to Louie Meyer and Dale Drake. Leo became chief development engineer for Meyer & Drake Engineering, the position he held for the nineteen years of that firm's existence. Company policy was very conservative and centered around steady refinement of the existing types of

Leo in 1954. The small supercharger impeller is from a Miller 91; the larger is from an Offy. GB

four-cylinder Offenhauser engine and constant improvement of the plant's production facilities. Thus, except for supercharged versions of the Offy Midget and 220, no novelties were produced by Meyer & Drake.

During this period, however, Leo worked long and hard in his "spare" time. In 1948 he designed one of the finest front-drive cars of all time for owner Gil Pearson. It was built but, due to an odd combination of events, it never raced. Then in 1956 there was the highly sophisticated supercharged 183-cubic-inch V8 which Leo designed for oilman Howard Keck. This machine showed every promise of being a world-beater, but for excellent personal reasons Keck had to shelve the project.

Then in 1959 came the Reventlow Formula-One *Scarab*. The engine's major characteristics were dictated to Leo, and he drew and detailed it, plus a 91-cubic-inch version which never saw the light of day.

When the DOHC Ford showed its superiority at Indianapolis in 1965 the world reacted as though something new had happened, as though a tradition had died. On the contrary, it was merely a case of carrying the same tradition a step further and is approximately what Meyer & Drake would have done had the tremendous financial resources been available.

In any event, the four-cam Ford V8 is a kissin' cousin to the four-cam Miller V8 of 1932. Leo served as a consultant when the Ford engine was being designed. The Ford's downdraft intake ports and four-valve combustion chambers are strikingly similar to those of that type which Leo laid out for the first time anywhere, back in 1929. And not only are the piston-type cam followers of the Miller-Offy type, Ford buys them from the specialists who build the Offy engine.

The Offy did not die on Memorial Day, 1965. There still was nothing to replace it in dirt-track racing. Meyer & Drake had been Ford's exclusive sales agent for the DOHC V8, and after the Indianapolis race Louie turned his interest in the firm over to his partner. He joined the Ford racing organization, and Dale renamed his company the Drake Engineering & Sales Corporation with, of course, Leo as its engineering chief. The big, 255-cubic-inch Offy was kept in production and a supercharged 170-cubic-inch version was added to the line. The many much more advanced engines that Leo had designed during the Meyer & Drake years remained on sheets of drawing paper that are tucked away in the Drake pattern loft.

Miller and Offenhauser together produced several engines and cars before Leo was absorbed into the team. Each had a certain class and quality that set it apart from its peers. With Goossen's entry a basic change took place in Miller's products. A new architectural mode began to take form and a new, much more professional and refined character began to emerge.

Young draftsman Goossen found the most fertile environment with Miller for what he had within him to express. This environment was composed of Miller's benign patronage, his boundless promotional daring, his volcanic imagination, and his confidence in Leo's dedicated ability to translate pencil scrawls and inspired mutterings into screaming and durable machinery.

It was composed, too, of dour but wryly humorous Fred Offenhauser's exceptional talents in commanding and efficiently administering what was, in effect, a combined jewelry and race-car factory. Miller lived for the dream of an unending series of ever more perfect machines. Leo and Fred provided the competent hands and feet-on-the-ground approach that enabled Miller to survive in the real world. What the three had in common was an acceptance of the concept of the thoroughbred machine—acceptance at the least and life-motive at the most. Each did his job, and Leo's certainly was not the least of them. Each learned from the other.

Boat-racing champion Ralph Snoddy said:

> Miller wasn't much of a mechanic himself but he had ideas, knew what he wanted and what should work. Like the little air pump we drove off the end of the camshaft to keep pressure up without constant manual pumping. He told Leo what he wanted and Leo drew it up. "No, that's not exactly it," Harry would say. "Turn this around a little bit and change this so." Leo would redraw it and Harry would say, "Yeah, *that's* what I want!" He couldn't do it himself.

Said the great Ed Winfield:

> If you want to get a racing engine designed Leo can do a top job for you in a tenth of the time it would take anyone else to do it. That's one of the big breaks Offenhauser and, later, Meyer & Drake had—getting the original Miller engineer.
>
> On the Novi project my brother Bud was project engineer, I was consultant and Bud was dependent upon Leo and me. Leo did all the drafting and most of the engineering. Using Leo, Bud and car owner Lou Welch did in a few months what otherwise would have taken a very long time and would have cost even more money. That he could do this showed that Leo had most of it in his head.

Van Ranst said:

> Leo really deserves the credit. When I came out to California to figure out the original Cord front drive I had to use the straight-eight Lycoming engine and we had to use the standard transmission parts. Well, I just couldn't make out of those junk components anything that I would consider building. It was Leo who took hold of the problem and said, "What difference does it make? We've got to build the thing." And Leo actually laid out most of that front end—for me, actually.

Gerald Kirchoff, practical engineer and one of Miller's most devout disciples, said:

> Leo was *the* brain behind the whole Miller-Offenhauser saga.

Said boat-racing champion Dick Loynes:

> Leo has always been remarkable for trying to help everybody and there never has been proper credit given to him and Fred. By giving freely of good advice and by playing ball with everyone they got the full cooperation of the racing fraternity, a vast fountainhead of ideas.

Said engineer and author Karl Ludvigsen:

> Throughout his career Leo probably never has been in a position to specify

the broad concept of an engine design: number of cylinders, ratio of stroke to bore—basic philosophy. These have been set by the sugar daddies and their promoter friends, who then turned to Leo to do the actual design. Very often his abilities and his fantastic experience have been able to make modest concepts look good (the immortal Offy) but they have not been able to save half-baked concepts (the Scarab) nor should they have been expected to.

I think that Leo has done an amazing job in view of the fact that he has never had the feedback that a designer gets when working for a firm that races the engines it makes. When his engines were built they left home forever, only to return when the owner wanted drawings for a revision which he had already preconceived. He has had to design, in short, without the feedback provided by development. He has done a heroic job.

Leo Goossen himself said:

Miller was a genius, but being without formal education he needed someone with my background to put his ideas into manufacturing form and to make them workable. Young as I was, it happened to be me; it seems to have been an act of Providence. Miller was most appreciative of my efforts and I remember when six 91 engines were being built and run through the shop at one time and my drawings weren't even looked at. Fortunately there were no errors and no changes were required, but you can imagine the responsibility and worry that I felt. Miller's confidence in my ability was an inspiration to do my best. That is how we all felt about working for him. That privilege was an indescribable reward.

It is remarkable how a whole vast discipline, cult, and tradition can be winnowed down to the point where so much of its essence exists only in the mind and memory of one solitary, quiet, gifted man.

PART THREE

The Beginners

chapter 9

DUESENBERG

WHEN MASON GAVE Fred and Augie the go-ahead to produce an engine for racing, most of the world's automotive engines were of the side-valve type. Whether they were L-head (a single row of valves) or T-head (a row of valves on each side of the cylinders, operated directly by two camshafts carried in the crankcase) their combustion chambers were large and rambling. Compression ratios therefore were inevitably low, and very slow crankshaft speeds were essential to give the poor fuel of the period a reasonable chance to burn.

Inherent too in these large combustion chambers were "dead pockets," areas where the products of combustion futilely tried to expand against immovable surfaces, instead of having their force concentrated over the head of the piston. And the water-jacketing, of course, was proportional to the size of the chambers, so that thermal efficiency suffered badly from heat escaping to the coolant instead of being transformed into kinetic energy. But the side-valve engine was wonderfully simple and cheap, and it had some mechanical advantages of its own. For one thing, a camshaft acting directly on side-valve tappets was every bit as positive and precise as an overhead camshaft acting on OHV tappets. As for the pushrod and rocker arm OHV alternative, it was a compromise that carried a number of penalties. Among them were lost motion, greater friction losses, the flexing of the pushrods, complexity, multiplicity of parts, and, relative to the side-valve approach at least, significantly higher cost.

In spite of their rustic isolation the Duesenberg brothers were sensitive to this

Eddie Rickenbacker at the wheel of Tom Alley's 1915 Indianapolis Duesenberg. IMS

problem. Perhaps their solution to it was original, but in any case it was thoroughly unorthodox and stands as a monument to their grasp of their art and to their courage.

They used a combustion chamber of the same diameter as the piston and with a vertical, rectangular recess over the head of the piston. Two horizontal valves, in cages, were screwed into each of these recesses, and opposite each valve was a magneto-fired spark plug.

The valves were operated by means of huge vertical rocker arms, about a foot long. Their lower ends bore directly on the cams and their upper ends directly on the valves. The system was cheap, simple, highly effective, and efficient. It was so fundamentally sound that variations on its theme were used by Delage in his 1914 Grand Prix engine, by Miller in his aircraft V12, and by Sir Harry Ricardo in his post-World War I tank engine. It survived in some Lycoming power plants into the late 1930's.

The Duesenberg walking-beam (so-called for the big rocker arms) engine had its four cylinders cast in a single iron block with integral head. The aluminum crankcase was of the barrel type and the crankshaft—installed from the flywheel end—ran in just two ball bearings. Winding out to 2,200 RPM and higher, the crankshaft would bow so much that the number two and three connecting-rod big ends would bang against the crankcase side-plates, in spite of an eighth of an inch of static clearance. This caused consistent bearing failure until the designers hit upon the offbeat idea of radiusing the lower-end bearings of the two central connecting rods, making them slightly concave. According to no less a Duesenberg authority than Karl Kizer, who knows these engines intimately, this expedient resulted in very acceptable reliability.

The lubrication system was of the dry sump type and had the impressive capacity of fourteen gallons, as insurance against squirting large amounts of oil on the track. After a brief bout with thermosyphon cooling, a proper centrifugal water pump was adopted.

Augie and his helpers made every possible part themselves, and their ingenuity was as marvelous as their industry. The big rocker arms, for example, they built up by hand out of flat sheet stock, shaping the metal, then riveting it together into an I-beam structure, and then welding on the fulcrum pivot, cam shoe, and adjustable tappet-screw. Eddie Miller recalled:

> When you think of the labor that we went through to make a part then, compared with conditions today, you can only laugh. But that was the way it had to be done because there was nothing available that would fit.
>
> We couldn't afford to buy our forgings so we made our own. Among other things, I did most of our blacksmithing. We hand-forged our own connecting rods and as long as we had single-plane crankshafts without counterweights we forged those. We forged our rear axle half-shafts and, later, wheel spindles and other intricate parts. We used Timken front axles and bought our frame rails from Parish, a specialist. Chrome-nickel steel was the best alloy of the time and that's what we used. Our cars were very strong. You could get into a terrific wreck and just bend one end or the other. The parts that stuck out usually were the only ones that suffered.

By 1912 Fred and Augie had their 4.316 by 6 inch, 350.5-cubic-inch engine delivering close to 100 HP and were ready to go racing with a vengeance, cheered on by their new patron, Frank Maytag.

The classic walking-beam engine. Intake "manifold" is between water pump and magneto mount. GB

Above left: Valves were removed through the spark-plug bosses and there were, of course, two plugs per cylinder. Note huge cone clutch and housing of cast aluminum. Above right: Front of engine with gear cover removed. Huge timing gear meshes with gear drive, left, for water pump, oil pump, and magneto. GB

They made a respectable debut that year in the Algonquin Hill Climb and by the end of the season were winning some important contests. They took first and second in the Wisconsin Trophy race at Milwaukee, a first in the Pabst Blue Ribbon Trophy on the same course, and finally a first and second in the *formula libre* race at Brighton Beach, N. Y., ahead of a field of excellent machines with vastly greater displacement. Fred and Augie's star had risen.

Their racing campaign was much more ambitious in 1913 and even more encouraging. Although their Mason cars were 100 cubic inches under the 450-cubic-inch prevailing formula, they entered and qualified four machines in the Indianapolis "500." Driver Willie Haupt came home in ninth place against the biggest and most advanced competition from home and abroad and ahead of a Peugeot, a couple of Mercers, and three Isotta Fraschini. Robert Evans' Mason was only thirteenth but the sharper observers began to watch "those Dutch farmers from Iowa."

The engine's huge vertical rockers and horizontal valves are exposed. In background is Karl Kizer, guiding light of the Indianapolis Speedway Museum. GB

The next year was one of the most intensive activity. Maytag began losing his interest in automobiles, and Fred and Augie cleared out of the backwoods. They moved to St. Paul, Minnesota, rented a few square feet of floor space in a large machine shop, formed the Duesenberg Motors Corporation, and began racing under their own name. At Indianapolis in 1914 Eddie Rickenbacker and Willie Haupt brought their renamed, rebored (4.4 by 6; 360.5-cubic-inch) Masons home in tenth and twelfth places. The first four finishers were Delages and Peugeots, but the Duesies outlasted eighteen good machines, including Boillot's Peugeot, Friedrich's Bugatti, and Chassagne's Sunbeam. Then on July 4 Rickenbacker won the big Sioux City 300-miler outright and Tom Alley came in fifth, well up in the money. The new house of Duesenberg raced the full season without other victories but still with a respectable record. That they did so with a 90-cubic-inch handicap told the knowledgeable that these farmers knew more about machinery than how to run a haybaler.

It was a feverish year, and Fred and Augie went where their cars went, Augie to keep them running and Fred to manage the team and look after business interests. Still, they found time to accomplish big things in the little shop in St. Paul. Big-time boat racer Commodore J. A. Pugh commissioned Duesenberg to build him a pair of engines to power his entry in the 1914 Harmsworth Trophy Race. The race was canceled because of the outbreak of the war but, with these engines, Pugh's *Disturber IV* was the first boat ever to break the mile-a-minute barrier.

Frugal Fred and Augie shrewdly set themselves up in the marine-engine business while executing their patron's order. The walking-beam engines they built for Pugh were in-line twelves—and one shudders to think how few main bearings they must have gotten by with. The cylinders were designed and cast in pairs so that, with the patterns and tooling for them, Duesenberg suddenly was able to introduce a line of two-, four-, six-, and eight-cylinder marine engines. Only the crankcases, crankshafts, camshafts, and a few other components had to be tailored to suit the demand and the demand proved to be brisk.

Up to this time the leading manufacturer of high-performance marine engines was the Loew Victor Engine Company of Chicago, headed by industrialist J. R. Harbeck, who also was a director of the American Can Company. The war had created a need for more powerful and efficient marine engines, and Duesenberg Motors had just what was needed to satisfy it. Harbeck sent Loew Victor engineer William R. Beckman to St. Paul with almost a blank check for manufacturing rights to the Duesenberg six- and eight-cylinder marine engines and for Fred's services as engineering consultant. Beckman completed his mission. Loew Victor swung into production of the Duesenberg Patrol Model Marine Engine and found ready markets for it with several governments, including those of the United States and Imperial Russia.

The big time was beckoning, and back in St. Paul Fred and Augie built a detuned prototype of their automotive engine which would be suitable for passenger-car use. In February of 1915 they announced to the press that these engines now were available to the industry. There were no immediate takers but their schedule was too full for that to matter much at the time. They raced a tremendously busy season, some trophies of which were Eddie O'Donnell's victory in the Glendale, California, road race in February, Ralph Mulford's win in the inaugural 300-miler on the new Des Moines Board Speedway in August, and O'Donnell's and Alley's fifth and eighth at Indianapolis. In the fall they moved into a much larger shop and also could afford the services of enthusiastic young engineer C. W. van Ranst.

In the spring of 1916 Van laid out a sixteen-valve version of the walking-beam four, simply adding a camshaft and a bank of rockers to the engine's other side and relocating the spark plugs lower in the head. This power unit went into a remarkably advanced car, built for Omar Toft and called the Omar Special. Toft wrecked it the first time out, and the engine's potential remained unknown for a while.

Shortly after this, Harbeck convinced Fred that with Fred's ideas and Loew Victor's production capacity they could go far together and Fred agreed to move his entire operation to the Chicago plant. Here the team built several new race cars and several sixteen-valve engines, and in August of 1916 completed the construction of one of the largest and most powerful aircraft engines that had been built anywhere up to that time. It was a 4.875 by 7 walking-beam V12, designed to deliver 300 HP at 1,400 RPM, and had its cylinders cast in pairs in the manner of the marine engine. It had a single camshaft at the vertex of its 60-degree cylinder banks and closely anticipated the Auburn Twelve that still was two decades in the future. It had structural weaknesses and too many other projects intervened for them ever to be corrected.

Still, 1916 was a good year, Duesenberg's most active and successful in racing up to that point. The season opened with Ed O'Donnell's victory in the important 300-mile Corona Grand Prize Road Race on April 8. A week later he won the inaugural 150-miler on the one-mile Ascot dirt track at Los Angeles. In the Indianapolis "300" of that year Wilbur d'Alene, with Eddie Miller riding beside him, finished a fine second in an eight-valve Duesey, less than two minutes behind Dario Resta's Peugeot. Then Tommy Milton took second place in the 300-miler at Tacoma, and d'Alene took second in the 300-miler at Cincinnati.

In time for the September Harvest Classic at Indianapolis Fred and Augie had prepared a team of three cars with 16-valve engines and with brand new planetary transmissions. As always, the cars were fired up for the first time at the very last minute—and they ran backwards. The new gearboxes were torn out, old ones reinstalled (both types were in unit with the rear axle), and d'Alene and George Buzane finished third and fourth in the main, 100-mile, event. Toft, in the rebuilt Omar Special, came in sixth, and it was clear that the 16-valve engine added another four to five MPH to the car's top speed.

The Duesenberg team contested every Championship event of the season, which closed with a Duesenberg-Rickenbacker victory in a 200-miler at Ascot. Rick finished third in National Championship point standings for the year, but many of his points were won at the wheel of a Maxwell (with SOHC engine designed by Ray Harroun).

But Wilbur d'Alene and Tommy Milton, driving Duesenbergs only, finished the season close to the top, in sixth and seventh places. Duesenberg was becoming a name to be reckoned with.

Its first success in the aeronautical field also came in 1916. Fred and Augie were approached by the Gallivet brothers of Norwich, Connecticut, who had designed and built an amphibious plane and were looking for the best means of propelling it. For this project Fred and his team concocted a pair of 5 by 7 bore-and-stroke, four-cylinder engines. They were mounted side-by-side, and pinions at the rear of their crankshafts meshed with the internal teeth of a 26-inch ring gear which drove a single huge four-blade pusher propeller. The arrangement bore a slight, strange similarity to the yet-unknown 16-cylinder Bugatti. Eddie Miller was sent to Norwich to make the installation, and he had the adventure of going on test runs. The pilot would fortify himself with a few belts of strong liquor, and they would take off, out to Fisher's Island, over the ocean, and back to the mainland. The big bird would put down in Norwich Harbor like a rock but it did fly, and its engines gave no trouble.

No sooner had Fred moved into the Loew Victor plant in Chicago than Harbeck proposed that they form a joint venture for the volume production of the walking-beam engine for the passenger-car industry and for the probable production of aircraft engines. Harbeck had the capital and the marketing contacts, and Fred had the technical assets on which a very prosperous business might be based. But Fred valued his independence, and it was months before he could bring himself to agree to the proposition.

The plan was that Loew Victor would provide Duesenberg Motors with a manufacturing plant. A site was acquired at Elizabeth, New Jersey, and construction of the plant was commenced in the fall of 1916. Simultaneously, and in order to get the program under way with minimum delay, Harbeck used his American Can Company connections to get Fred and his team installed temporarily in the ACCO plant at nearby Edgewater, New Jersey. The team included shop foreman Jimmy Lee, who had been with Fred and Augie since the Des Moines days and even before, in Mason City, Iowa. There were also Eddie Miller, Ernie Olson, C. W. van Ranst, and, as the team's first full-time draftsman, Bill Beckman. He was backed up by George Dennis; and then there was Tommy Milton, as at home in all the engineering arts as he was at the wheel of a winning race car.

They worked in a dark cramped corner of the ACCO factory, but Fred converted a bedroom in his home into a drafting office in which some wonderful projects were born. One was the original Duesenberg automotive straight eight, work on which began toward the end of 1916. It still utilized the walking-beam principle. But there was little time to think of automobiles now; the wartime need for aircraft engines was growing more desperate by the day.

The state of the art was primitive indeed. There wasn't much to fly. Wright got started with a two-cylinder training engine designed by Charles Lawrence. It was air-cooled, vibrated like mad, developed only about 20 HP, but served to teach bud-

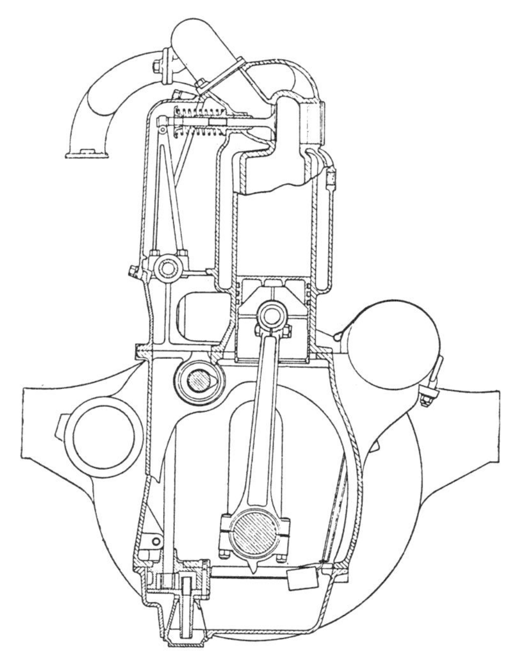

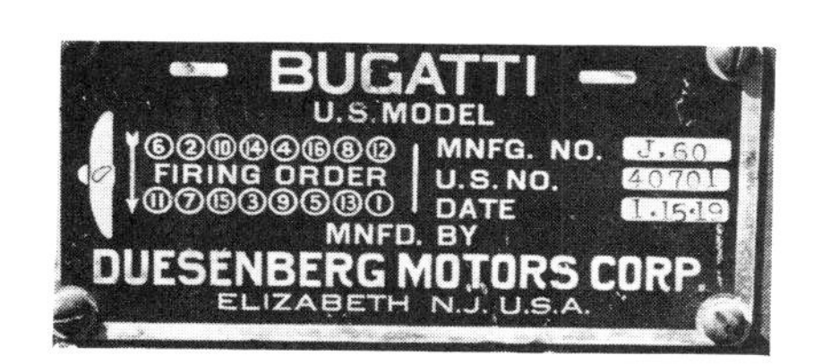

Above left: The V12 aero engine on its primitive test stand. EDDIE MILLER

Above right: Transverse cross-section of Rochester-Duesenberg engine. GB COLLECTION

Left: Bugatti "parallel sixteen" nameplate, from an engine presented to the author by Richard Zanteson. GB

Below: The gigantic V16 aero engine. EDDIE MILLER

ding pilots how to taxi a plane. Curtiss built the OX-4, a little V8 that pulled only about 65 HP. Then came their highly touted OX-5, which never developed enough power or reliability to go to war. Wright, before mastering the radial-cylinder configuration and introducing the J-5, built Hispano-Suiza engines under license at Trenton, New Jersey. They were excellent but were good for only about 220 HP in their largest form. And there was the Liberty engine, the first one of which was tested by the United States Bureau of Standards on July 4, 1917. It was made in V8 and V12 form, the latter being good for about 360 steady HP or 400 for perhaps a minute. So there was a need for aircraft engines over a wide specification range.

It was during the ACCO period that van Ranst drew up a 4.75 by 7-inch version of the 16-valve racing engine, geared down 1.73 to 1 with aircraft use in mind. It weighed less than 500 pounds, developed 125 HP at 2,100 RPM, passed the government tests, and was adopted as a power plant for training craft.

Duesenberg Motors moved into its Elizabeth headquarters early in 1917. The "plant" turned out to be very small and poorly equipped. Ernie Olson recalled the little milling machine that bounced around on the floor every time it made a cut, and the crude dynamometer which used wooden paddles slapping the air for engine loading. At no time did the "plant" boast more than ten men on its staff, but they were brave men and true, and they accomplished wonders.

The little artisan workshop continued to crank out training-plane engines and made such an impression upon the War Department that it was included in the original program as one of the firms that would manufacture the Liberty engine. But eventually someone in authority visited the "plant" and decided that manufacturing giants such as Packard, Ford, Lincoln, and Marmon were better equipped to perform the miracle of producing 15,000 Liberty engines in eighteen months. Duesenberg Motors was given contracts that were more in harmony with its hand methods and its record of achievement in the areas of research and development. It was chosen to build the 16-cylinder Bugatti aero engine and also to design and build an engine for bombers and perhaps dirigibles—something really big and powerful, more so than any aircraft engine that ever had been built in the United States.

Just as many British writers have made it a practice to credit Henry with the creation of the *Charlatan* Peugeots, so have they magnified the significance of the Bugatti straight eight, perhaps beginning with a retrospective essay by E. W. Sisman in *The Automobile Engineer* for July, 1927.

True, Bugatti produced a straight eight in 1913 and then, early in World War I, designed an aircraft engine which consisted of two of these, with two crankshafts on a single crankcase. The engine was manufactured in France by Bara of Levallois, where Henry was on the engineering staff. It was made in the United States by Duesenberg. Henry then designed the straight-eight Ballot, and Duesenberg appeared with his somewhat similar engine. But it is a romantic oversimplification to make Bugatti the father of the idea. This impression was fostered by the coincidence—not at all odd in view of the specialties of these men—that Duesenberg and Henry each appeared with his own in-line eight simultaneously and immediately after having worked with Bugatti eights.

The idea was at least as old as the French CGV of 1902; as old as the Premier *Comet* of 1903; or the Winton *Bullet No 2.* of 1904; or the British Weigal of 1907; or the early Mercedes Zeppelin engines, to name only a few pioneers of the in-line eight.

Perhaps Henry, indeed, was inspired to adopt eight cylinders for his Ballots as a result of the Bugatti experience, but his engines nevertheless were still just stretched Peugeots. What has been overlooked in the case of Duesenberg is that this firm's Patrol-Model Marine Engine was in volume production in straight-eight form by Loew Victor in 1914 and was being sold to various of the world's navies. Fred and Augie recognized its advantages then, which is why they hastened to lay out an automotive engine along the same lines. It could not be built until after the war, when war-busy private enterprise could return to private pursuits. But it could be and was built to meet military needs—in gigantic and sophisticated V16 form.

The advantages of the in-line eight versus the four or six were listed in Sisman's paper:

1—High ratio of mean to maximum torque.
2—The reciprocating parts are in both primary and secondary balance.
3—More efficient cooling than an engine with a lesser number of cylinders for a given piston displacement.
4—Decreased stresses in the working parts for a given piston displacement and a given speed.

The Bugatti twin eight which was intended to profit from these advantages had three vertical valves per cylinder and a shaft-and-bevel driven SOHC for each of the two vertical, parallel banks. The two crankshafts were geared together and the propeller hub doubled as a cannon which spat buckshot through its 1.46-inch bore. This was an idea borrowed from Hispano-Suiza.

Ettore Bugatti had built the prototype for this engine at his factory at Strasbourg in Alsace, which was then German territory. Very early in the war he transferred his headquarters to Paris, taking the engine with him, via Italy. There he and his old friend W. F. Bradley were in constant touch, and it was Bradley who arranged for Bugatti to present his engine to the technical committee of the Bolling Mission.

The Mission was named for its director, Colonel R. C. Bolling, a New York attorney and son-in-law of President Wilson. It was part of the supply service of the United States Army, and its task was to go to Europe, build up aircraft squadrons out of whatever equipment was available, train American personnel in their use, and send them to the front. One of the Mission's main assembly and training bases was where Orly Airport now stands, on the outskirts of Paris.

Bradley had been doing work in Europe for the American Committee of National

Defense, and within days of Bolling's arrival in Paris, Bradley found himself a member of the Mission with the rank of captain; his encyclopedic knowledge of the European automotive and aircraft industries was invaluable.

When the search began for European aircraft engines that could be manufactured in the United States, it was conducted by the Mission through a team of dollar-a-year experts which included Colonels J. G. Vincent, chief engineer of Packard, Howard Marmon of Marmon, and E. J. Hall of Hall-Scott. It was Vincent and Hall who later designed the Liberty engine, and Hall was to have a subtle but vast influence on American racing engine design. It was this group which, after witnessing the Bugatti engine under test, voted for its acceptance and then chose Duesenberg to build it. Bradley recalled:

> While they were testing that first engine there was an American soldier working in the shops. He stepped in front of the propeller and was killed—the first American soldier to be killed in Europe. We decided that the engine and two mechanics from the Strasbourg factory—who were officially German—should be sent to America, and also the body of this first of the war dead. Our transport officer applied to the French authorities for a permit for one corpse, two Germans, and an aviation engine. You can imagine the reaction. When they calmed down enough to speak they told our man very gently, assuming that he had lost his mind, that what he asked was quite impossible. After reminding them that Napoleon had decreed that there was no such word he finally got his permit.

The engine failed miserably in its first tests at McCook Field (later Wright Field), and pioneer automotive engineer Charles B. King was sent to the Duesenberg plant to redesign the engine and make it work. On February 23, 1918, the first of the King-Bugattis was tested at Elizabeth, in the presence of a large assemblage of dignitaries. Van Ranst recalled:

> I was in the test room when the first engine blew up. Our excuse for a dynamometer was just rocking stands and a club in place of a propeller. When the engine let go all that was left of it were the mounting brackets. The club went throught the roof and was found several hundred feet away. It was a miracle that no one was injured, much less killed.

Milton recalled that there never was one of these engines that would run for more than four or five hours, nor was there one that ever got into the air, although forty had been delivered to the government when the contract was cancelled.

The major fault—there were myriad smaller ones—was this: the engine had been designed with optimum compactness in mind. It was very narrow, and its crankshafts were timed to mesh with each other exactly like the rotors in a Roots blower. The torsional vibrations in the crankshafts were tremendous and resulted in the rapid wear and breakdown of the gears that united them and drove the propeller. Of course the instant that the two crankshafts tangled at speed the engine was reduced to a heap of rubble.

The sixteen-cylinder engine which the Duesenberg team designed and built while this drama was going on owed absolutely *nothing* to Bugatti influence. It used three valves per cylinder but there were one intake and two exhausts—the reverse of Bugatti practice. And of course the valves were actuated by walking beams and a single camshaft just as the V12 of 1916 had been. The highly sophisticated Duesenberg V16 was a monument to weight-saving and followed Mercedes practice in having fully machined cylinder sleeves and heads with welded-on water jacketing of thin, stamped sheet steel. Bore and stroke were 6 by 7.5 inches, and total piston displacement was 3,393 cubic inches, yet, complete with carburetors and magnetos, the engine weighed only 1,250 pounds. Swinging a gigantic sixteen-foot propeller, it developed about 800 HP at 1,250 RPM, and did so with reliability. But the war ended and, with it, any need for such a gigantic and costly power plant.

On the subject of engineering influences upon the Duesenberg team, the following points are significant. As already noted, the Wright Engine Works was building Hispano-Suiza SOHC V8 engines for the government, as Duesenberg Motors was building the Bugatti for the government. There was the freest possible exchange of technical information between the two plants, which were just fifty miles apart.

Then, after the Bugatti-engine explosion at Elizabeth, the government equipped the plant with a motor-generator dynamometer of the very best quality and used the plant as a test facility. This dynamometer could absorb better than 1,000 HP and it could be used to "motor" engines electrically. Thus, during the last half of 1918 the Duesenberg crew had the opportunity to work with and observe closely practically every high-output engine that was of interest to the United States Government.

Finally, America's entry into the war in 1917 did not put an end to racing. The Indianapolis Speedway property was immediately turned into a military aircraft repair depot and a landing field and Carl Fisher went off to Florida to begin the transformation of Miami Beach from swamp into what it is today. But racing continued on the boards and dirt, and Duesenbergs, usually the "factory team," again took part in every race of the year. On May 30, in the 250-miler at Uniontown, Milton and Hearne were fourth and fifth out of a field of twenty-seven. On June 16 Hearne was fourth in the 250-miler at Chicago out of an equally large field. Milton and Hearne were second and fourth in the Omaha 150-miler on July 4. The 100-miler at Providence on September 15 was won by Milton, Hearne coming in third. Then, in the last two races of the season, the 170-miler at Uniontown and a 50-miler at Ascot, Hearne and Milton finished one-two in both events. They won fourth and fifth places in National Championship point standings.

There was little big-time racing in 1918 but much other business went on as usual. Early in the year the Revere car went into production, powered by the Duesenberg walking-beam four-cylinder engine. Within months the Roamer and Wolverine appeared, offering this engine in 85 and 103 HP versions. Harbeck and Fred were realizing their dream, and the cars that could boast "the power of the hour" ranked as the purebred aristocrats of the hour.

In 1919 Tommy Milton installed a modified training-plane engine in a Duesenberg chassis and in November at Sheepshead Bay set new absolute records

Above: The walking-beam, straight-eight marine engine also was made in four- and six-cylinder versions. JERRY GEBBY

Left: A most historic picture, showing a block for the walking-beam eight-passenger car, with two blocks for the SOHC racing engine in the background. Taken at Elizabeth, N.J., plant. JERRY GEBBY

Below: The walking-beam eight that originally was to power the Model-A passenger car. JERRY GEBBY

for everything from one to 300 miles and from one to three hours. Although his engine had less than 300 cubic inches' displacement he averaged an unheard-of 116.2 MPH for 25 miles.

Soon eleven different makes were getting their light, efficient engines from Duesenberg. Their excellence was underlined on April 22, 1921, when a Roamer roadster set a new AAA Stock Chassis record of 105.1 MPH at Daytona Beach.

But Fred and Augie had foreseen another future when they projected their walking-beam automotive straight eight in 1916, seeing in it the power for both a racing and a passenger car. The wartime experience altered their thinking to the extent that they came to favor a single overhead camshaft, driven by shaft-and-bevel gears. As the tide of the conflict turned and Allied victory seemed imminent, the lights burned seven days and nights a week over the wall-to-wall drawing board in Fred's upstairs front bedroom on Frelinghuysen Avenue in Elizabeth, N. J. Thus, like Ballot, they produced a team of radical new race cars for the first postwar Indianapolis "500," which took place less than six months after the Armistice.

Duesenberg's preparation, however, was not quite as thorough as Ballot's. Two of the old four-cylinder cars were more than ready but the 300-cubic-inch straight eights were barely assembled at the final moment. Charger Tommy Milton drove his car to the track in time, but the other two new entries became hopelessly trapped in race-morning traffic and never reached the starting line. Milton lay contentedly in the middle of the pack until a connecting rod let go on Lap 49. But the Ballots were shockingly fast as long as they were running well.

They vindicated Fred's faith in the straight-eight principle. He forged ahead with it, and in December of 1920 he exhibited the prototype of the Duesenberg Model A passenger car at the New York Salon. It was fitted with the late-flowering walking-beam straight eight but the public was advised that the production model would have a SOHC engine.

A bright new future lay ahead of Fred Duesenberg and, to pursue it, he rid himself of his rights in the walking-beam engine which he realized must soon yield to the overhead camshaft principle. In late 1919 he resolved his affairs with Harbeck, and the Duesenberg four became the property of John North Willys' Rochester Motors Corporation. The team remained in Elizabeth for a little over a year to assist in the program that adapted the Rochester-Duesenberg to mass production and then, in 1921, moved to its final headquarters in Indianapolis. It had not been an easy life for Frederic Samuel Duesenberg but, by thinking and working and fighting with the energy of an army of men, he had arrived at a point where he was slightly wealthy and where the promise of the future was dazzling.

NOTE: Considerable pains have been taken in this chapter to undo old misconceptions concerning the effective primacy of Bugatti's exploitation of the straight-eight principle. The major motive for this has been to set the historic record straight. A secondary motive has been to correct the author's own acceptance and dissemination of a myth which, merely by managing to survive, had become an apparent part of the substance of history.

chapter 10

FRONTENAC

LOUIS AND ARTHUR CHEVROLET WERE ABSORBED into the Durant-General Motors organization in 1907. Buick already was a part of it, was highly active in racing, and it was for their talents in this area that the brothers' services were retained. In 1909 the Buick *Bug* cars made their racing debut. There were two of them, and they were campaigned by Lewis Strang, and Louis and Bob Burman. In his car on June 19 Louis won the 395-mile Cobe Trophy road race at Crown Point, Indiana. Exactly two months later Louis won a ten-mile preliminary race on the inaugural day of the new, then macadamized Indianapolis Motor Speedway. In the big, 250-mile Prest-O-Lite Trophy race he led against the cream of the nation's cars and drivers for more than half the distance. But the track surface had begun to break up early in the event, and as Louis was overtaking another car on Lap 52 a flying stone shattered one of his goggles' lenses, sending splinters of glass into his eye. Temporarily blinded and in great pain he managed to get back to the pits, with his mechanic's help, and then to medical aid. Burman, in the other *Bug*, took over the lead and won the race, as he and Louis were to do many times again.

In the almost nonexistent literature on the life and work of Louis Chevrolet there is a tendency to credit him with the design of every piece of machinery with which his name ever was connected, and the *Bug* is no exception. In this case, while both Louis and Burman were urged to advise as racing experts, the project was under the supervision of Buick engineer A. E. de Waters. The actual design layouts were

Louis Chevrolet at the wheel of what was probably the world's first monocoque race car and also the first car with four-wheel independent suspension. HOWARD BLOOD

made by E. C. Richards, a graduate engineer of French descent who did a major portion of the detailing of most of the early Buick cars. But Leo Goossen never forgot the imposing figure of Louis Chevrolet on his advisory missions to the engineering department.

Louis' ideas must have been good, and he must have demonstrated good executive ability for Durant, in addition to securing the use of his name, to contract with Louis for the design of the original Chevrolet car and for construction of a prototype. Together with his powerful patron Louis spelled out the general specifications. But the actual engineering and detailing were done by another forgotten man, Etienne Planche.

Who was Planche? Planche was a young Swiss or French draftsman who, in about 1906, found work with the Walter Automobile Company (today the Walter Motor Truck Company) of New York City, then the motor capital of the western hemisphere. The firm's founder was a Swiss, and Louis also worked for him for a spell as a mechanic. It was a small company, and it is perhaps here that Louis and Planche first met.

Needing larger facilities Walter built a plant at Trenton, N. J., to which Planche was transferred. By 1908 he had the official title of General Manager and Designer for the Walter Company. In that year he designed a jewel of an air-cooled in-line four-cylinder motorcycle engine which was marketed by another Trenton firm. Then a Walter offshoot was the Roebling-Planche car of 1909 (three models, the biggest having 120 HP). And then, early in 1910, the Mercer Automobile Company flowered from these sources.

Planche's role in the creation of the immortal T-head Mercer (the Raceabout was the very first model, being the simplest) is not clear, particularly since C. G. Roebling and Finley R. Porter both outranked him in the firm's engineering hierarchy. He left Mercer and joined Louis in Detroit and created the first Chevrolet car.

When Louis broke with Durant in 1914 he immediately founded his own Frontenac Motor Corporation, with Albert Champion's backing. The new firm's initial purpose was to build racing cars, but these, undoubtedly, were to be merely a bridge to the manufacture of passenger cars. Still with Planche doing his engineering, Louis soon had designs for two different ultra-modern engines and a complete car. Brother Arthur had two new machines under construction when Louis had his final and near-fatal showdown with Champion, which is why the 1914 Frontenacs did not race until 1916: Louis had to find new funds. On the strength of his work for Chevrolet and Frontenac, Planche soon became chief engineer of the Dort Motor Car Company. He held this position until 1923, when he set up his own consulting offices in the General Motors building in Detroit.

To pick up the next thread in Louis' story we must flash back to 1912 when Howard Blood, of the Blood Brothers Manufacturing Company of Kalamazoo, Michigan, had designed the most radical of production-car prototypes. He called the car the Cornelian, for the blood-red gem stone, and entered a specially prepared version in the 1915 Indianapolis race. Said *Motor World* of May 26:

CORNELIAN AN ODD BIRD

A particularly interesting small car because it is so small is the Cornelian. This grasshopper among the elephants has a . . . motor of a paltry 116 cu. in displacement and yet can lap at over 80 MPH owing to the negligible weight and windage of the tiny machine.

It has no frame at all, as the sheet metal body acts as a frame as well, and still more extraordinary, it has no axles. This sounds an impossibility, but is explained by the spring system which replaces the axle proper. At the rear there are three springs, placed transversely one above and two below, and the extremities of these springs are linked to brackets which carry the rear road wheels. The transmission is fixed to the frame-body and drives to the road wheels by two universal shafts after the fashion of the old de Dion cars.

For the front axle, a similar construction is used, only without the driving shafts, of course (and only one upper and one lower spring), and the car is steered by link rods and chains attached to a small drum which is keyed to the foot of the steering column.

In 1919 Ralph Mulford's aluminum-engine Fronty qualified at Indianapolis at 100.5 MPH. GB COLLECTION

Of course the car is not expected to come in a winner, but if it completes the course it will be a vindication of many original principles for light car design.

So, a half century before Clark, Chapman, and Lotus arrived at Indianapolis, there was a streamlined, superlight, monocoque car running there with fully independent suspension on all four wheels. The contemporary writer cited the de Dion system because it was the closest approach he knew of to truly independent suspension of the driving wheels. The Cornelian's compact transmission-differential unit was mounted on the monocoque structure, making it sprung weight.

Blood built a few cars for sale during 1913 and early 1914, but both the inboard and outboard universal joints on their rear axle half-shafts were made of leather, the weak point of the system. Then Blood replaced the leather with thin, tempered-steel strips, resolved his one remaining major problem with it, and prepared to put the Cornelian into mass production in mid-1915.

On September 26, 1914, a talent-packed hundred-mile race was held on the Kalamazoo mile dirt track. At the last minute Blood had the idea to strip the fenders from one of his little, less than 1,000 pound roadsters, head for the track, enter the car, and look for a driver. Twenty minutes before the start "Cap" Kennedy gave up finding a proper ride and settled for the Cornelian. The remarkable thing is that, never having seen or touched the car before, he finished seventh in a field of thirteen huge, hairy monsters, all at least three times the size of the Cornelian, including Burman's Peugeot, de Palma's Mercedes, a Grand Prix Delage, and two Duesenbergs.

This amazed a good many people, including Howard Blood. He decided that if he could do in the Indianapolis 500-miler what he had done at Kalamazoo he should have no trouble selling 2,500 Cornelians at 410 dollars the copy (top and windshield 25 dollars extra) by the end of the year. But he needed help of the most competent kind.

The Cornelian used a four-cylinder, 18 BHP, pushrod, 2.875 by 4, 95-cubic-inch power unit made by the Sterling Engine Company. This firm was one of the almost innumerable parts of Durant's newest empire and was managed by W. H. Little, who had been one of the principal figures in the forming of the original Chevrolet Company and previously had been Buick's plant manager. It was probably Little who counseled Blood that Louis was just the man to mastermind the entire Indianapolis caper and who brought the two men together.

Louis needed the job and took it. He welded a tapered tail to the roadster body, mounted oversized wheels, reworked the cylinder head to take two Master (Miller) carburetors, bored and no doubt stroked the tiny engine out to its 116 cubic inches, and did everything else he could to make the car raceworthy, including signing Joe Boyer on as relief driver.

Louis was so well prepared that the Cornelian was the first car to qualify for the 1915 "500"; in fact, it was the only car to take the trials on the opening day. Its four-lap average was 81.1 MPH—just behind a 300-cubic-inch Bugatti at 81.5. In the race itself things went well enough for Louis, aside from a little overheating and running out of gas on the backstretch, until the engine swallowed a valve at 180 miles. The car retired and was instantly forgotten by the world.

If monocoque construction and independent suspension had such great potentialities why were they apparently lost on Louis Chevrolet? The only advantage of monocoque construction for the one-off or few-off builder was weight-saving, and Louis' Frontenacs soon would teach the world a lesson or two about that. Independent suspension might be good for road racing, but road racing had just died. The Speedway era was just beginning, and it would take fifty years to prove that solid axles aren't necessarily the ideal for paved oval tracks.

And the fate of the prophetic litle Cornelian? Production was dropped at the end of September, 1915, after only about 100 cars had been built. The Blood brothers were doing so well as U-joint manufacturers that they had no time left for their former underpriced sanguine hope.

By hook and by crook Louis, Arthur, and Gaston finished three of the Frontenac

With flat radiator Louis qualified his aluminum-engine Fronty at 103.10, second only to Thomas' Ballot. IMS

race cars in time for the 1916 Indianapolis "500." Arthur qualified one car at a tolerable 87.70 MPH. Arthur qualified another at only 82.45, whereupon its crankshaft broke. Gaston could not squeeze the required 80 MPH out of his mount, but Joe Boyer managed to do so and Arthur co-drove this car in the race. Arthur went out on Lap 35 with ignition failure. Louis lasted until Lap 82, when he was eliminated because of a broken connecting rod. It was hardly an encouraging start. He had to skip most of the season's many races but did patch up a car for the 300-miler on the Cincinnati boards in September. He was merely one of the twenty-two non-finishers out of twenty-eight starters. He missed the next seven big races of the season but was busy ironing out the kinks in his machines. Then, in December, the 112.5-mile inaugural race on the Uniontown Board Speedway was won out of the blue by Louis Chevrolet at the searing average of 102 MPH and the Frontenac legend was on its way.

The "iron engine" which Louis and Etienne Planche had developed followed GP Peugeot practice in general configuration but differed in many details. It had a cast-iron block, integral cylinder head, twin overhead camshafts driven by a train of spur gears, finger-type cam followers, and an aluminum crankcase which was split down its horizontal center line. Louis built two of these engines, but their power output and weight were disappointing and they were shelved. Ironically, this was an inherently excellent design, and it became the basis for the 1920 Indianapolis winner.

Satisfied that he was getting nowhere trying to beat Peugeot at their own game, Louis went to work with Planche on a new approach in the fall of 1915. It was undoubtedly the Mercedes victory in the 1914 French Grand Prix that inspired Louis to build an inclined OHV, SOHC engine, but it was even less derivative than the iron engine. It was well documented by the press because it employed aluminum to a totally unprecedented extent, to an extent that had been considered impossible. Taking the car as a whole, these are some of the parts that were made of aluminum: cylinder block, pistons, crankcase, intake manifold, camshaft cover, transmission housing, gear covers, water pump, oil pump, clutch cone, rear axle center section, brake flanges, pedal brackets, starting crank, body panels, and underpan. All this in spite of the fact that aluminum was in its infancy, its alloys were porous and poor, its metallurgy was practically nonexistent, and its welding techniques were a scarcely known black art.

The four-cylinder aluminum engine had a bore and stroke of 3.870 by 6.375 inches. The single overhead camshaft was driven by a vertical shaft and bevel gears and operated four valves per cylinder, two 2-inch inlets, and two 1.75-inch exhausts. The valves were actuated by roller-type rocker arms, and each set of four rockers was carried in its own aluminum housing, in which it could be efficiently lubricated and adjusted. Each valve worked against two concentric coil springs whose diverse characteristics were calculated to cancel out harmonics. So much for the Mercedes influence.

The cylinder head, block, and crankcase were one integral casting, making it a foundryman's nightmare. A pretty touch that added to the complexity was that all four valve seats for each cylinder consisted of a single sort of cloverleaf iron casting, around which the aluminum itself was cast. Huge sideplates in the cylinder-block water jacketing probably were downright necessary in locating the cores. Thin-walled, cast-iron dry cylinder liners were shrunken into the preheated block.

In most other details the iron and the aluminum engines were identical. Like the 1913 GP Peugeot, both had ruggedly dimensioned crankshafts which ran in three large ball races and were bolted together at the central main bearing for this purpose.

The dry-sump lubrication system of both engines was quite advanced for its time and used both feed and scavenging pumps. A twelve-gallon reservoir was mounted on the front of the firewall, its surface covered with copper fins for heat radiation. The underpan and internal sheet-metal ducting were contoured so that all air passing through the radiator had to flow over these fins before escaping. Two independent oil systems were fitted, each with its own pumps, one to act as a standby in case of failure of the other. And in case these precautions should not be enough, the riding mechanic was provided with a hand pump with which the system could be pressurized in case of emergency. An interesting detail of this system was the delivery of oil under low pressure to the main bearings, after which it was picked up in hollow rings which utilize centrifugal force in feeding the oil to the connecting rod bearings, from which oil spray reached the cylinder walls and wristpin bearings. This was another idea from the 1913 GP Peugeot, and Duesenberg was to adopt it in 1920.

Both engines rode on four tubular mounts and were entirely separate from the transmission, which had its own three-point mounting amidships. This made for a very rigid layout, an open flywheel, and further weight saving. Hotchkiss drive was used, also in the interest of weight reduction, and in these cars Louis used a differential-less locked rear end, which was no novelty at the time. Suspension was by conventional semi-elliptic springs.

Louis' handling of the fuel storage problem was admirable. He provided two long, narrow gas tanks below the frame and toward the rear on either side of the car. His reasons were the logical ones: equal weight distribution, concentration of weight toward the driving wheels, and lowering the vehicle's center of gravity.

The Frontenac body was praised for its clean, graceful, aerodynamic lines, which even its underpan shared. And Louis provided leaf-spring-spoke steering wheels for the reduction of driver fatigue.

Louis built four of the aluminum-engined cars, and they had a ready-to-go dry weight of about 1,750 pounds, a good 660 pounds under the GP Mercedes, at least 500 pounds under the lightest American competition, and even about 40 pounds under the 183-cubic-inch 1913 GP Peugeot.

After the debacle at Indianapolis in 1916 Louis took a position with the American Motors Company of Plainfield, N. J., manufacturer of the American Beauty Six. He was in charge of development and testing, and each car bore a nameplate that testified, "OK—Louis Chevrolet."

Throughout 1917 and 1918 the Frontenac team contested every National Cham-

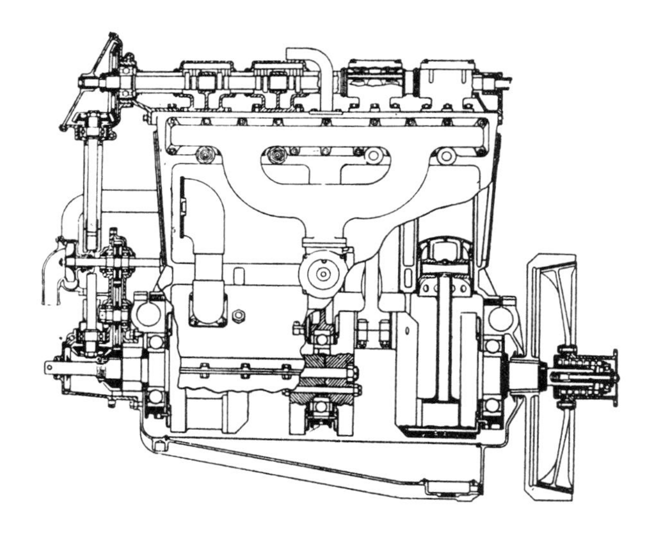

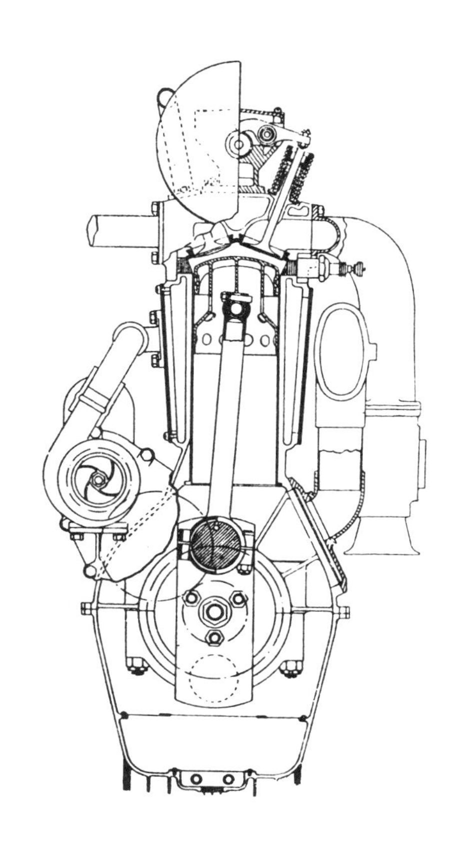

Right: Transverse cross-section of the aluminum engine. GB COLLECTION

Above: Longitudinal cross-section of the aluminum engine. GB COLLECTION

pionship event; there were plenty of them the first year, and the entry lists were huge. In the Uniontown 300-miler on May 10, 1917, Boyer was second and Louis fourth. In the 250-miler at Cincinnati on May 30, it was Louis first, again, and Gaston third. Nothing then until a 100-miler at Chicago on September 3, won by Louis. On September 15 at the Providence, Rhode Island, one-mile cement speedway, Fronties were first, second, and fourth in the five-lap opener, then did poorly in the following events. At Sheepshead Bay a week later Louis finally had all four of his cars running at once. That 100-miler was won by Louis at a brilliant 110.4 MPH; Ralph Mulford's Fronty was fourth, and those of Joe Boyer and Dario Resta were dead last. Next came Chicago on October 13. In the 10-lap opener Gaston and Mulford took second and third. In the two 25-milers that followed, Mulford won the first and finished second in the second. The season closed at Ascot on November 29; Louis won the 10-miler and finished third in the 50-miler. He finished that season second in National Championship, just a handful of points behind Earl Cooper.

The United States was so deeply involved in the war in 1918 that there were only two races of major consequence. The first, a 100-miler at Sheepshead Bay on June 1, was won by de Palma in a twelve-cylinder Packard. The second, for the same distance, at Chicago on June 26, was won by Louis at 108.12 MPH.

Racing quickly returned to normal in 1919 and, as tired as they deserved to be, there was still terrific performance in the four old Fronties. The 100-miler at Sheepshead Bay on July 4 was won by Gaston at 110.5 MPH. On the same day, 3,000 miles to the west, Louis won an 80-miler at Tacoma. On September 1 Boyer won a 225-miler at Uniontown and then on September 20 Gaston won the 150-miler at Sheepshead Bay, this time averaging only 109.5 MPH. But for sheer drama Indianapolis was the high point. The four old 300-cubic-inch Fronties were qualified by Louis, Gaston, Mulford, and Boyer. Boyer went out on Lap 13 with a rear wheel adrift. Seven laps later Mulford left the race with a broken driveshaft. But Gaston finished a not-ignoble ninth and, said *Motor Age:*

> The real race was between Louis Chevrolet and de Palma for sixth and seventh places. Lap after lap they fought, de Palma coming up from the ruck after changing a wheel bearing and Chevrolet fighting to make up the time lost in changing a steering knuckle, tie-bar and wheel. The battle brought the grandstands to their feet time after time—this after the race had been won. So close was the finish that only a re-check of the timing tape could determine who got in first. If de Palma or Chevrolet, who between them led the field for the first half of the race, could have kept up their pace, the track record for the distance would have been broken.

And it was a grim race, the bloodiest in the Speedway's history to that point. Louis who, between 1905 and 1920, lost four riding mechanics and spent a good three years recovering from crashes, became involved in this day's misfortunes too. Near the end of the race he came hurtling down the grandstand straight and lost a front wheel. With marvelous skill he kept the car perfectly under control as its wheel spindle ground over the bricks. But as he crossed the start-finish line the spindle snapped the wire which lay across the track and triggered the Speedway's timing and scoring equipment. The broken wire flew back and severed an artery in the neck of E. T. Shannon, who was preparing to overtake Louis. Hardly knowing what had happened, Shannon blazed on, but when his mechanic brought him into the pits at the end of that lap Shannon was in a state of collapse from loss of blood. He survived, his mechanic finished the race, and thanks to the whole freakish incident, the Speedway devised a new means for tripping its clocks.

It was during Louis' wartime stint with American Motors that he met van Ranst, who was working at the nearby Continuous Casting Corporation on a pet project of his own, a 300-cubic-inch six-cylinder DOHC *supercharged* aircraft engine. This was the combination, Louis and Van, that would give all the postwar greatness to the Frontenac name.

chapter 11

MILLER

IT WAS ABOUT 1907 that Harry Miller set up his own business in Los Angeles. His shop was just a backyard lean-to which he managed to equip with an old lathe and drill press. Here, about 1909, he built the first Master carburetor, his invention. A Mr. Beasley came forward with financing, and operations were moved to a small factory on Los Angeles Street. Sales and production boomed, and in 1911 Harry could afford to buy his first professionally manufactured car—a Lozier, one of *the* thoroughbreds of its day. That May he went to Indianapolis for the first time. He practically lived in the Speedway garage area, observing everything, meeting everybody who was anybody in racing, and explaining to each of them the virtues of the Master curburetor.

Miller never lost his interest in the challenge of the problems of carburetion as long as he lived, and he continued to design and build carburetors until at least the late 1930's. And of course this deep interest and the years of specialization that it had produced were basic to Harry's swift mastery of the art of supercharging in the mid-1920's.

His Master *Automatic* was an ingenious approach to the then-universal problem of providing proper fuel-air mixtures for a wide range of engines and over their entire RPM ranges. The Master performed brilliantly on race cars, police cars, fire engines, and other vehicles that were called upon to do their best under full-throttle conditions. At part throttle its limitations became more marked as competing

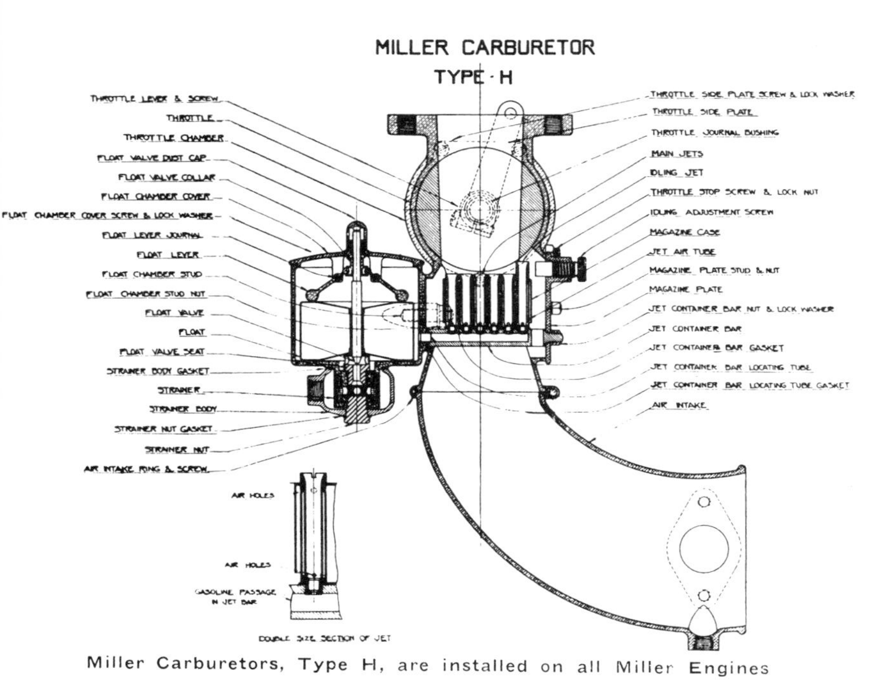

Miller's carburetors, on which his early success was founded, were notable for their batteries of jets that were progressively uncovered by a barrel-type throttle valve. GB COLLECTION

carburetors became more refined, and the Master finally vanished from the market at about 1921.

The two most distinctive features of this carburetor were the "jet bar" and the barrel or rotary throttle valve with its odd spiral opening. The jet bar was a horizontal tube which was mounted parallel to the axis of, and beneath, the throttle valve. Screwed into the tube were eleven to nineteen tall, vertical jets. Due to the spiral opening in the rotary throttle valve, when the valve was closed only one jet remained open, for idling requirements. The wider the throttle was opened the more jets were uncovered, and vice versa. It was easy to change jet sizes and, while the Master was quite inefficient at part load, hardly anything could match it for racing. Which is why Harry Miller's carburetor factory happened to become the West Coast hangout for most of the American racing fraternity.

Fred Offenhauser joined Miller in 1913, just as he was branching out into the manufacture of aluminum pistons. His seems to have been the first firm in the United States and perhaps anywhere to mass-produce pistons in light alloy, and this brought Miller still more racing trade.

Offenhauser was born in Los Angeles on February 11, 1888, of German parents. He was one of the best machinists and toolmakers in the finest machine shop on the West Coast at that time—that of the Pacific Electric Railway. He heard that Miller needed a toolmaker and paid good wages.

> I went down and talked with Miller and told him that I wasn't interested in making a change if I had to be somebody's relative to get ahead. I wanted to get ahead on my own ability. He seemed to like that and he hired me.

Fred did get ahead fast, and soon Miller asked him to take over as plant superintendent. He agreed, but only on the condition that there would be one boss over the shop force—himself. Soon Miller's shop had the reputation of being the cleanest and most efficiently run on the Coast. Miller never interfered in Fred's decision when he hired or fired, nor did he ever ask why. Their relationship grew into seventeen years of very ideal understanding, harmony, and friendship. And after that, of course, it was Fred who carried on the Miller tradition.

On one of his Indianapolis trips Miller had met the son of Charles W. Fairbanks, Teddy Roosevelt's vice president. Young Fairbanks was greatly impressed by the Master carburetor and told Miller he was losing fortunes by not distributing his product nationally. Fairbanks became so convinced that he bought the Master assets from Miller and Beasley late in 1914 and set up a large factory in Indianapolis, where he manufactured these carburetors in large volume for several years. Miller's output had been in excess of 5,000 units per month for the previous couple of years.

Well compensated by Fairbanks and probably drawing royalties from him, Miller designed and began to manufacture the Miller carburetor under the egis of the new Miller Carburetor Company.

This carburetor was designed purely for racing use and was not remotely in competition with the Master. It was a rudimentary and simplified form of the Master which embodied an almost total lack of concern with part-throttle operation. It was precisely what was needed for the just-commencing speedway era. The spiral opening was dispensed with in the rotary throttle valve and was replaced by a simple circular bore. The jet bar was retained but the number of jets was greatly reduced, and the bar was mounted at right angles to the throttle-valve axis so as to provide an at least theoretical progressive uncovering of the jets. These few elements plus a float bowl constituted *the* American racing carburetor until the advent of the superior Winfields in the mid-1920's.

All this tended to make Miller's plant racing headquarters for an ever-increasing segment of the fraternity. The segment became nearly total as the word got around that Offenhauser could fix or build anything on a racing car. Thus everything came to Miller's for repair: Isotta, Fiat, Benz, Delage, Mercedes, as well as most of the domestic makes; and none of these educational chances of a lifetime were wasted on Harry and Fred.

It was in the winter of 1914 that they got their first taste of actually building an engine. Bob Burman, then billed as the King of Speed, had blown up his Grand Prix Peugeot. He needed a new engine for the racing season ahead and cabled the French factory to that effect. Peugeot replied that they were at war and no longer in the racing business, so Burman took his problem to Miller.

There was little that could be salvaged from the 1913 GP Peugeot. The crankcase, crankshaft, cylinder block, connecting rods, and pistons, all were rubble. The most important parts that could be used again were the camshafts and their housings. A slight additional problem was that all the Peugeot's dimensions were metric and would have to be translated. And Burman had to have in his hands, within four months, not just another engine, but one capable of out-performing the best opposition in the world.

Offenhauser analyzed the problem for a couple of days, then told Miller that he could do it, but only by working several men day and night, seven days a week, until the very last minute. Miller passed on this information and an estimated fee of 4,000 dollars to Burman, who gave the go-ahead. While they were at it and with their client's approval, Miller, Offenhauser, and chief draftsman John Edwards did all they could think of to improve upon the Peugeot design. They increased valve and port sizes and improved the port contours. The Peugeot H-section connecting rods were a known weak point of the design and were replaced by superbly finished tubular hand-forgings—what were to become typical Miller-Offy rods. Instead of machining the new pistons from steel billets *à la* Peugeot, Miller, of course, used his "Alloyanum" aluminum alloy pistons with a single ring each. The four pistons weighed 44 ounces less than the originals, and a considerably greater saving of weight was realized through the new light but very strong rods. Then, Burman's 4.5-liter, 3.7 by 6.9 inch, 274-cubic-inch Peugeot did not take full advantage of the 300-cubic-inch displacement limit. The new, Miller-built engine had a bore and stroke of 3.6 by 7.1 inches, for a displacement of 298 cubic inches.

Finally, one photo of the Burman engine is known to exist; it was reproduced on page 22 of *Motor Age* on May 27, 1915. The article which accompanies it states that the rebuilt engine's valve gear is standard Peugeot, but one glance at the photo shows that Burman and Miller evidently had adopted a single overhead camshaft, again, probably being under 1914 Grand Prix Mercedes influence. This entire job was so critical and so much was at stake that Offenhauser did most of the machine work himself. The deadline was met, and after his first track test of the new engine Burman, delighted, estimated that he was at least ten HP better off than before.

Burman campaigned this car constantly during the busiest-ever season of 1915, which was dominated by Dario Resta and his own 4.5-liter Peugeot. But on May 1 Burman won the 200-mile Southwest Sweepstakes Road Race at Oklahoma City. At Indianapolis he qualified in seventh place and finished sixth. His only other win that season was a 25-miler during the inaugural events on the one-mile asphalt speedway at Providence, Rhode Island. Then he began preparing for the big 1916 opener, the 300-mile Corona Grand Prize Road Race near Los Angeles.

Over 100,000 spectators turned out to see one of the last big races on the open road, and the fastest of them all. At 268 miles Burman was lying in second place, hoping to seize first in the final few laps, when a tire blew as he was lapping at 97 MPH. The blue Peugeot flew out of control; it slammed against a utility pole, then a parked car, then another pole as Burman and riding mechanic Eric Schraeder were hurled 50 feet through the air to their death. Burman actually expired in his wife's arms. It was more than strange that two years before, at the previous Corona Grand Prize, she had had a terrifying dream which she had confided to a friend. The friend told a journalist, and the press was filled with the story. It turned out to be a very accurate description of Burman's fatal crash.

One of Miller's most important carburetor clients was the Hall-Scott Motor

Company, and it seems to have been Colonel Hall who referred Lincoln Beachey to Miller while the Burman job was in progress. It is almost impossible to imagine today that such a short time ago there was just *one* man who was *the* aviator in the United States. But there was just one, and he was Beachey. He had made a great deal of money as a barnstorming pilot and had made such a name that he had little difficulty in raising a large amount of capital when he decided to go into the business of manufacturing aircraft.

Beachey roughed out his general engine requirements for Miller who consulted with Offenhauser and John Edwards and came up with a detailed design which Beachey approved. He urged Miller to complete a reliable prototype as soon as possible for Beachey to demonstrate at the World's Fair—the Panama Pacific Exposition—in San Francisco, which was to open on February 20, 1915.

Beachey had specified that the engine have six cylinders. There were two valves per cylinder in pent-roof chambers and actuated by a shaft-and-bevel SOHC. A pair of Bosch magnetos provided current for the dual ignition system and Miller would hear of nothing but dry-sump lubrication. In the interest of very generous main-bearing support the aluminum crankcase was split several inches below the crankshaft centerline. The camshaft, crankshaft, and cylinder barrels all were machined from steel billets, and the engine's general architecture had much in common with Mercedes aircraft practice. Like Mercedes, the new six had built-up water-jacketing, but with a typical Miller twist. The steel cylinders were mounted in pairs on a jig, and beeswax was molded around them where the water-jacketing should be. Then the entire assembly was placed in an electroplating bath, and about .047 inch of copper was applied over the beeswax. Then the wax was melted out and the job was done. Experts said that it would be impossible but the plating procedure actually was successful. But it also was slow and expensive and lacked rigidity and was replaced with more conventional, welded-on copper sheet.

The prototype engine ran gratifyingly well, and Offenhauser went north with it to Beachey's small plant at Redwood City. There he installed it in a two-place biplane and went on many shakedown flights with the daredevil pilot. When the combination was working perfectly Beachey cut aerial capers over the heads of the World's Fair crowds that made him the darling of the public, the press, and of the financial backers he hoped to win. They underwrote his manufacturing project, but on the major condition that he give up flying and devote himself to running the business. But Beachey made just one last flight to test some new controls, crashed into the San Francisco Bay, and the whole project died with him. Only one of these engines was built.

But one job led to another. Early in 1916 a physician who was also an aviation enthusiast came to Miller a request for a high-performance, light-weight aircraft engine. Almost simultaneously A. Cadwell, sponsored by Bell & Howell, asked Miller to build him a complete race car. Other orders soon followed, resulting in the first six Miller four-cylinder engines. Two went into airplanes and the remainder into cars.

Miller showed his preoccupation with esthetic effect from the start, and the new

Six of these four-cylinder, SOHC, largely aluminum engines were built by Miller in 1916—two for aircraft, four for race cars. TED WILSON

engine presented a clean exterior with all possible organs enclosed in aluminum housings. Breaking with the Beachey engine and its Mercedes ideas Miller adopted:

1—Barrel-type crankcase, giving 360-degree support to the main bearings.

2—Integral crankcase and cylinder-block casting in aluminum, using wet cylinder liners of machined steel.

3—Camshaft drive by train of spur gears.

4—Four valves per cylinder, operated by a single overhead camshaft, and rocker arms working directly against the valve stems. Huge port areas.

5—Detachable cylinder head with rubber rings to seal the wet cylinder liners.

6—Two-piece crankshaft with single ball race at center main bearing, double ball race at front, and two double ball races at rear main bearing.

7—Tubular connecting rods; cover-plates and hand-holes on sides of crankcase for easy, quick removal of rods and pistons.

All six of these engines had a bore of 3.625 inches. The two aircraft engines had a stroke of 4.125 for a displacement of 170 cubic inches. The automotive engines, designed to compete under the 300-cubic-inch formula, had a 7.0-inch stroke and a displacement of 289 cubic inches.

The Cadwell machine was beginning to take shape when Barney Oldfield saw it and decided to order a car for himself. He had confidence in Harry Miller and his organization after they had done a masterful job of rebuilding his Delage. That car was over the hill now, but it had one feature that Barney had learned to believe in: desmodromic valves. He and Miller conferred, and the result was an almost entirely new engine. The dry-sump barrel-type crankcase was retained, but a cast-iron block with integral head was adopted, with gear-train-driven overhead camshafts and, of course, valves which were closed mechanically and were not solely dependent upon all-too-fragile and fallible springs. Output of 125 HP at 2,950 RPM was admitted, but Offenhauser stated years later that the engine was cammed to peak at 4,000 RPM, which was tremendously high for the times.

What was most remarkable about the car was its body—the most fully streamlined vehicle ever seen in the United States. It seems to have been conceived jointly by Miller and Oldfield and was radically ahead of its time to say the least; its shape in plan view was an almost perfect teardrop. Almost two decades before Carrozzeria Touring in Italy was to develop its celebrated *Superleggera* structural technique it had been anticipated in this body, which was formed of arc-welded aluminum sheet over a light steel framework. And, interestingly enough, the cockpit was equipped with a roll bar and was claimed to be "crash proof." Curved glass or plastic were not available, and the window openings were covered with wire screen, but the wind was so nicely deflected by the contour of the cowl that Barney claimed he could drive the car at full speed without goggles and enjoy a cigar at the same time. The other Miller cars of this period had equally fine bodywork, but of the open, two-place type.

Work on the Oldfield machine began in the spring of 1916, and it was a full year before the owner took delivery and handed over the last of its 15,000-dollar price. Finished in mirrorlike gold lacquer the car was promptly named the *Golden Submarine,* and it was a tremendous crowd attraction because of its bizarre and beautiful appearance if for nothing else.

In its first outing at the Chicago Board Speedway on June 16, 1917, Barney cut some promising practice laps at 102 MPH, but in the race itself his engine gave out after only ten miles and after he had averaged 104 MPH up to that point. Earl Cooper's Stutz won the 250-miler at an average of 103.15. But during the next week Oldfield cured his engine troubles to the extent that on the Milwaukee Fair Grounds mile dirt track on June 25 he defeated his arch enemy, de Palma, in each of three match races. Ralph's big Packard *Twin Six* was a highly developed and reliable machine, and this performance taught some respect to the scoffers who had rechristened Barney's folly the *Deviled Egg* and the *Golden Lemon.*

But that was just the beginning. Six times during 1917 Oldfield and de Palma dueled on the dirt, and four out of those six times "The Barber," as Oldfield called his rival, lost. At St. Louis on August 9 Barney proved that his was the fastest dirt-track combination in the world when he ran against AAA clocks and broke every international record for dirt tracks from one to 50 miles. His average for the mile was 80 MPH, and for 50 miles it was 73.5.

Anyone who has driven a latter-day Italian competition coupé at speed knows that the internal noise is infernal. Barney must have had his fill of the car when, late in the season, he flipped the Sub into an infield pond at Atlanta and barely escaped drowning. So, the car began the 1918 season with a conventional open cockpit. Barney raced it with good success, and at Uniontown in May be trounced Louis Chevrolet in a Fronty with an average of 105 MPH. Then he retired from racing, sold the car, invested heavily in the Firestone Tire & Rubber Company, and proceeded personally to put Firestone into the racing rubber once and for all.

All four of the original Miller four-cylinder race cars started out in life in the 1917 season. They ran in seven Championship events that year but did quite poorly except for a second place by Gil Anderson behind Mulford's Fronty in a 50-miler at Chicago. It is believed that two of the desmodromic iron engines were built, and Anderson's probably was the mate to Oldfield's, since the others seemed to have no speed at all, as Tommy Milton notes in Chapter 13. Even Cliff Durant, Frank Elliott, Roscoe Sarles, and fearless Leon Duray could not make them go.

After the death of Lincoln Beachey, De Lloyd "Dutch" Thompson became America's most glamorous aviator and a very wealthy man. He approached Miller in 1917 with the proposal that they pool their talents in the development and promotion of a large and powerful military aircraft engine which, if successful, could make them both immensely rich. He offered Miller a check for 50,000 dollars just to get things off the ground, and a partnership was born.

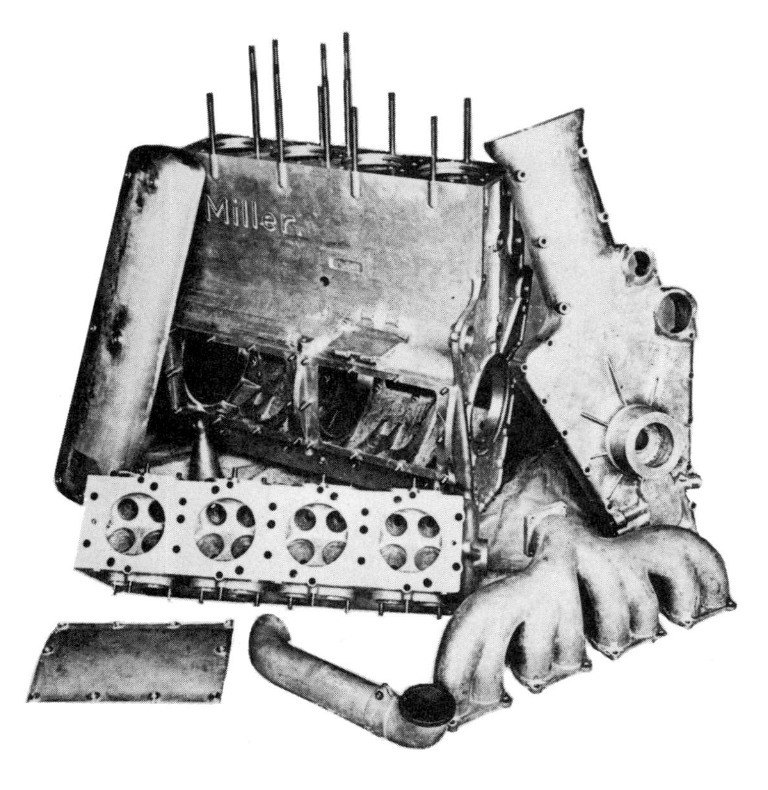

The first Miller four used wet steel liners, gear-train camshaft drive, and four valves per cylinder. TED WILSON

Right to left: Mechanic Claude French, Harry Miller, and sponsor A. Cadwell with the first complete Miller race car. TED WILSON

As usual, Miller came up with the unusual. It was a 5 by 6 inch bore and stroke, 1,414-cubic-inch V12, and again, a remarkably clean design. The crankcase and the two cylinder blocks were a single aluminum casting—the largest that had ever been attempted in the United States—and two efforts failed before a usable casting came out of the sand. This engine was strikingly similar to the V12 which Duesenberg had built a year before and which had received considerable publicity. It was a horizontal-valve, walking-beam engine, its most original feature being the operation of each pair of opposite valves by a single cam and a single forked rocker. This and the bizarre cam contour show clearly in an accompanying illustration.

This engine endured a 100-hour test, swinging a huge propeller at Miller's plant, and its output was estimated to be about 500 HP. Then, said Milton:

> They took it to Dayton Field for government testing and Miller walked out on it, as he did with so many other projects. He left "Dutch" flat with this engine. For all anyone knows it might have been a good one.

In this instance Miller's loss of interest was influenced by a contract to manufacture carburetors and fuel pumps in New York for the King-Bugatti engine which Duesenberg was building. Miller set up a small plant on 64th Street near Broadway in May, 1917, and was joined by Offenhauser in August. The little Miller Products Company was just getting around to running well and profitably when peace broke out, the contract was canceled, and Harry and Fred returned to Los Angeles in February of 1919.

Business continued as usual at the Miller Carburetor Company for a couple of months, when the owner of one of California's largest breweries backed Miller in the development of a radical, all-out race car which, however, with slight modifications, would be suited to volume production as a sports model.

This project was just getting organized, the name TNT chosen, and the general engine configuration agreed upon when Leo Goossen walked into Miller's office for the first time.

The TNT engine, as Miller had conceived it, had a barrel-type aluminum crankcase with integral cylinder block, wet liners and detachable cast-iron head. There were four valves per cylinder, two springs per valve, a pair of chain-driven overhead

The *Golden Submarine,* commissioned and campaigned by Barney Oldfield, had some notable racing successes and brought Miller his first fame as a car builder. TED WILSON

camshafts, and cup-type cam followers. The most remarkable feature of this engine was that its external surfaces were so clean and smooth that the car it was to power would require no hood; the engine would blend into the radiator and body panels.

Goossen detailed it, a prototype engine was built, and a TNT Special was entered for the 1920 "500" with Frank Elliott as driver. And then the project was dropped "for financial reasons." It served, however, as a brief rehearsal for the teamwork, the straight eight, and the thoroughbred period that all were about to begin.

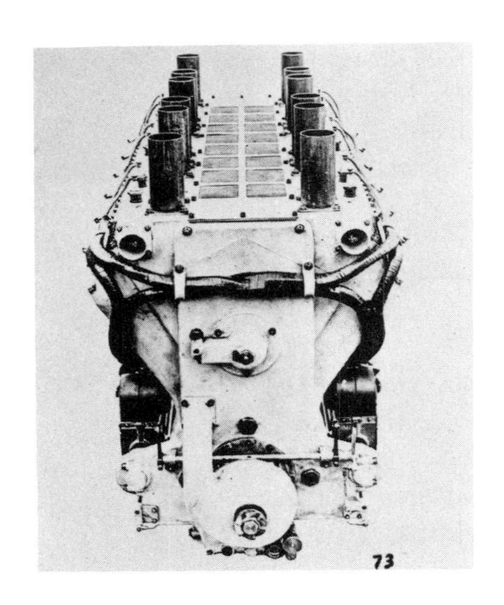

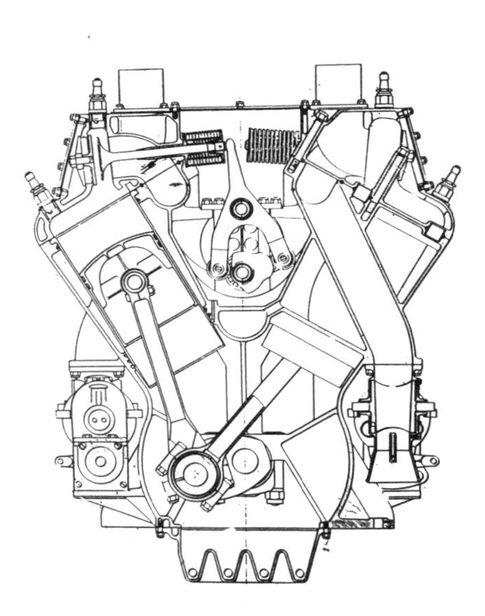

Left: Miller's V12 aero engine of 1917 had a 1,414-cubic-inch displacement. Its block was the largest aluminum casting ever attempted in the United States. Right: Transverse cross-section of V12 suggests Duesenberg walking-beam influence. Note shape of cam lobe. TED WILSON

PART FOUR

Fulfillment: The Roaring Twenties

chapter 12

DUESENBERG: THE YEARS OF GLORY

THANKS TO THE wartime 16-valve training-plane engine Duesenberg was one of the most promising contenders in the first postwar Indianapolis race, the last to be run under the 300-cubic-inch limit. The 1919 racing version of this engine had a bore and stroke of 3.75 by 6.75 inches—299.5 cubic inches—and five cars used it. They were Eddie O'Donnell's Duesenberg, Kurt Hitke's and Louis LeCocq's Roamer Specials, Wilbur d'Alene's Duesenberg-Shannon Special, and Arthur Thurman's Thurman Special.

There was no doubt that all these cars were extremely fast—much faster than their inexperienced drivers, as the press flatly pointed out. But the race was a debacle for Duesenberg. On Lap 27 Thurman's car flipped, killing the driver and almost killing the riding mechanic. On Lap 56 Hitke retired with a burned-out bearing. Four laps later piston failure eliminated O'Donnell. Then on Lap 97 LeCocq's fuel tank exploded, he lost control, the car overturned, and he and his mechanic were cremated beneath it. The last survivor of this doomed legion was d'Alene, who joined the spectators when an axle shaft snapped on Lap 120.

It was no consolation to Fred that he was negotiating to sell his rights to the walking-beam engine anyway. This disastrous public spectacle was a severe blow to his name, product, and bargaining position.

1920's sharply turned-out team had as drivers Hearne, Milton, Murphy, and O'Donnell. GB COLLECTION

But that was not all. The new eight, on which Fred and his little team had staked everything they had—blood, sweat, tears, and coin—was a dismal failure too. Three of these cars were built at Elizabeth. They were completed so late that there was no time for practice at the Speedway but only for qualifying attempts. They washed out their lower-end bearings, and the one car which Milton managed to qualify dropped out from the same cause on Lap 49 of the race. Due to their late completion and very negative showing these cars were ignored almost totally by press and public. But Ernest Henry's Ballot eights fared a little better, showed at least the potentiality of very superior speed, and sowed the seeds of faith in the straight-eight principle. Eddie Miller recalled:

> Our first 300 cubic inch eights had a splash-feed lubricating system, just like the fours. The fours never gave any trouble if you kept the oil level in the crankcase high enough and we had an adjustment in the cockpit whereby the driver or mécanicien could keep it regulated. Of course we used the same satisfactory system for the eights. But it took only about four laps, turning close to an unprecedented 4,000 RPM at the Speedway, for the connecting rods to start coming out. That's what happened the first year we ran the 300 eights—the rods all came out. Our latest cars weren't worth a darn.

So we went back to Elizabeth and we made a fixture. We had two-piece cranks that were put together at the center main bearing by means of a taper. The rear and center mains were ball races that ran quite happily in oil vapor from the engine and the front main was a plain bushing that you shoved the crank into.

We took a crank apart and drilled its whole length. We fitted a sort of tubular connecting rod at the front to feed oil under pressure to the hole in the front main. And we made a rig to wind the crank out to racing speeds in order to find out where the oil would go and at what pressure. We found that it took 60 PSI even to get it to come out of the rear hole, due to internal resistance and going over centers so many times. That was how we learned that we had to run in the 80 to 120 PSI range to be sure to get Number Eight main bearing lubricated. Then we began to wind those things like mad.

It was during this same period of desperately intensive lubrication studies that the Duesenberg team made one of its most historic discoveries. The standard connecting-rod bearings of the period were heavy bronze shells which were lined with a very thick coating of babbitt. Tests at the new racing RPM's not only pinpointed the melting point of the white metal but also showed that it took only about another 50 degrees F. for the bronze to melt and fall into the crankcase. So the team, recognizing the need to transfer heat from the bearings more efficiently than ever in the past, decided to dispense with the ponderous inserts, to bore the connecting rods out only about .040 inch larger than the crank journals, and to pour the babbitt directly onto the rods. This seems to have been the beginning of the thin-wall bearing, and the technique soon was applied to main bearings as well, all to the total incredulity of the Europeans when they heard of it. This system improved heat transfer vastly, and the production-car industry quickly began to adopt it. It trebled bearing life in the 300-inch Duesey eights, and by the end of 1919 and the five-liter 300-cubic-inch formula, the engine's lower end was quite refined. The top end still was running by guess and by God.

After the Indianapolis rout Tommy Milton rescued the honor of the walking-beam four by winning the historic, 301-mile Elgin Road Race in August. Then in November Milton, still hobbling on crutches after having almost lost his life at Uniontown nine weeks before, set his nineteen new AAA records in the 300-cubic-inch class with a 16-valve walking-beam four. During the same two-week period Murphy and O'Donnell broke all records in the 161- to 183-cubic-inch class from one to 300 miles and from one to three hours. Then, in the 301- to 450-cubic-inch class Dave Lewis set all new records from one to 100 miles and for one hour. Milton's speeds were the best, with 117 MPH for one mile and 113 for one hour. In all, 52 new records were established during this session at Sheepshead Bay, and the stock of the walking-beam engine soared to a new high. Fred got a good price from Rochester for the four-banger and hurled all his formidable energies into his vision of the future of the "vertical eight" and the passenger-car empire which he foresaw for it.

Among the historic incidents of this period was the first encounter between Wade

Murphy and Olson won the opening event on the Beverly Hills boards in 1920 with 103 MPH for 250 miles. GB COLLECTION

Eddie Miller and mechanic Pedro Nielson finished fourth at Indianapolis in 1921; tail section disintegrated. GB COLLECTION

Morton and Eddie Miller. Morton was vice president and general manager of the firm, which embraced the Biddle Motor Car Company and Meteor Motors Incorporated of Philadelphia, both of which used Duesenberg four-cylinder engines. Morton promoted these passenger cars through racing, first with a 16-valve four in a Meteor Special. Then he fitted this chassis with one of the new Duesenberg eights (probably one of the early walking-beam eights) and the car was campaigned in 1919 by Dave Lewis (Edna Miller's brother) with Eddie Miller (no relation to Harry) as riding mechanic. Eddie and Morton developed a tremendous regard for each other's abilities, and in the late Twenties they became E. L. Cord's two-man racing department. It was single-handedly responsible for all of the remarkable speed achievements of the Auburn car.

The Duesenberg engine for the three-liter, 183-cubic-inch formula of 1920 through 1922 started out simply as a scaled-down version of the 300, including its single overhead camshaft and three valves per cylinder. But getting a few of these 2.5 by 4.625 eights built in time for the first "500" under the new displacement limit was quite a drama.

Fred relied upon a patternmaker and foundry in Chicago for his castings, even though his own plant was about 1,000 miles away. This was not unusual; Miller had to go 500 miles to Berkeley, California (where the McCauley Foundry did Hall-Scott's work), to find castings of uniformly high quality. So, just a month before the 1920 Indianapolis race, all the engine castings for the new 183 Duesey were ready to be shipped from Chicago, when a freight strike was called. Fred rushed Eddie Miller to the Windy City with orders to return immediately with the iron even if it meant lugging it home on his back. The situation was utterly desperate.

In Chicago Eddie made a rendezvous with Ernie Olson, who was on his way to Elizabeth to help build a Duesey 183 for Ira Vail's Philbrin Special, the new name he gave the Thurman car when he bought the wreck. Miller and Olson descended upon the foundry with a supply of old Army foot lockers which they bought for 6 dollars each in a local surplus store. Into these they packed the cylinder blocks, heads, and other pieces for five engines. They had reserved a compartment on the first train to Jersey and they slipped the crew a 50-dollar bribe to let them carry their "hand luggage" on board. They sat on top of the 1,100 pounds of raw castings all the way to Newark, and when they reached Elizabeth it was Memorial Day—race day—minus three weeks! Somehow the engines were built and installed in their chassis in the time available, but even so the four cars had to be shipped to Indianapolis without axles. These were still being machined, and when they reached the Speedway and were assembled on the cars there was no time for practice and barely time for qualifying. But this was normal. Wilbur Shaw remembered, "Usually the Duesenbergs ran out of time as well as money. More factory-built Duesenbergs have shown up without paint at the Indianapolis Speedway than all other makes of cars combined."

In spite of such overwhelming odds the new 183's ran well. The engines pulled a few more horsepower than had been hoped for, and they enabled Milton, Murphy, and Hearne to finish third, fourth, and sixth in the "500." The team followed the Championship Trail for the rest of the season and did tolerably well. On June 19 Milton won the 225-miler at Uniontown at 94.9 MPH, and the season ended with a victory by Murphy in the inaugural 250-miler on the new Fresno boards. Still, the machinery was terribly folkloric, and mechanics in need of a piece of bailing wire would ask for "a Duesenberg hose clamp." Eddie Miller said:

> When he got going, Harry Miller's stuff was as fine as any that's ever been made. It was a joy to work on because his parts always were interchangeable from one engine to another. Every part was drawn and jigs were made for every operation.
>
> But with Duesenberg no two things were the same. The combustion chambers all were different. Some valves would stick up an eighth or even a quarter of an inch higher than they were supposed to. When the rocker arms came down to

meet them they'd bend the rockers away. The machining of the distance from the valve spring seat to the valve keeper might give you twenty pounds less pressure than the man who had the cylinder head that was made just before or after yours. He'd be two or three miles an hour faster than you and yet you both had the "same" parts. They could get away with this to some extent with the old four-bangers but of course the faster things go the greater the need for precision becomes.

Although it was going fast enough to make a good showing and to win an occasional race the Duesey 183 still hurt for power. But this fact was overshadowed by a much more acute problem: the lack of simple reliability. It was impossible to keep valve springs in the engines.

Fred had been in contact with Colonel E. J. Hall in connection with various aircraft engine projects. Hall was fascinated by the technical challenges of racing and was a cherished source of engineering counsel to the fraternity. Fred wrote him concerning this problem; and Hall replied that if Fred would make an engine available to him he would see what he could do.

Those were the free and easy days when you could drive an open-wheeled, open-exhaust race car on California roads and city streets with impunity and without license plates. Fred's cars were on the West Coast, and Olson was dispatched from Los Angeles to Berkeley at the wheel of one of the 183's. He reveled in the adventure, even though it included a violent storm on the then very primitive and dangerous Ridge Route.

Colonel Hall designed a test rig for the Duesenberg valve gear which was capable of putting it through even half-cycle runs under simulated racing loads. He found that the cam-lobe accelerations were so violent that springs could be broken with a single opening as fast as they could be installed.

Hall then went to work on the design of new cam contours that would be more suited to the very high RPM's that the 183 was turning. The essential characteristics of the new cams were an easy entrance curve to overcome the inertia of rest with acceptable accelerations, then a very steep rise, then an easy nose and a quick drop to a gentle cushion. This new design did not just put a virtual end to valve spring breakage. It also enabled Olson, who had driven north with 98 BHP, to drive back south with 115. It put another easy ten MPH into the cars and transformed them into real front-line contenders.

But the Hall camshaft was not developed until late in 1921. Without it the only win of the season was Eddie Hearne's 110.2 MPH average for the Cotati Board Speedway's 150-mile inaugural meet on August 15, following a 2-4-6-8 finish at Indianapolis by Roscoe Sarles, Jimmy Murphy, Eddie Miller, and Bennie Hill. With it, Hearne averaged 109.7 MPH for 250 miles to win the closing event of the season at Beverly Hills. And then there was a little mid-season victory overseas: Murphy in the French Grand Prix at Le Mans. It was that, the first-ever win by an American driver and/or machine in a major European event, that put the superlative, "It's a Duesey," into American slang and culture.

The 183 was running beautifully now, and the 1922 season began with Milton winning the 250-miler at Beverly Hills in March at 110.8 MPH. Then the next two Championship events fell to Murphy: victory at Fresno followed by a fabulous 114.2 MPH for 100 miles at Cotati. His car was a Duesenberg, but its engine was not.

After his Le Mans victory Murphy had purchased the winning machine from Fred. Then, the Hall camshaft had no sooner demonstrated its excellence than Miller got the design by devious means, and his own 183-cubic-inch eight began to fly. Murphy was one of the first to hear the news and to buy a Miller engine and install it in his Duesey chassis. With this combination he won the 1922 Indianapolis "500" at 94.48 MPH, breaking de Palma's 89.84 Mercedes record which had stood since 1915.

While the winning chassis was a Duesenberg and seven of the first ten finishers were Duesenbergs throughout, including the mount of second-place Harry Hartz, the whole balance of power in the racing world changed overnight. Suddenly it was Miller who had "the power of the hour," and clients flocked to him. From this point on, Duesenberg was eternally on the defensive, trying, with only occasional success, to keep up with "those damned cowboys out West."

The most ironic thing about the situation was that it had been Murphy who had

Left: At Beverly Hills in November 1921, Hearne's winning GP Duesenberg averaged 109.7 for 250 miles. Note front brakes. Right: Winner of the 1921 French GP now shines in the Indianapolis Speedway Museum. GB COLLECTION

defected to the other camp, had revealed Miller's might to the whole watching world, and triggered the westward stampede. It was no easy thing to break into the narrow ranks of racing drivers and if, through the faith of one's patron, one reached the heights, a certain loyalty might be expected. Beyond that, the charming and capable Murphy had been treated like a member of the family by Fred and Mickey, his wife. They never forgot this "betrayal." Milton, who felt just as wronged by that trio, viewed its breakup as just retribution for what he felt it had done to him. "I *made* Harry Miller," he was fond of saying. That was an overstatement, an indirect way of saying that Fred got what he asked for.

Partisanship became intense and remained so for decades. Art Pillsbury said:

> There are people around who will tell you that there never was a Miller engine, that Miller was just a carburetor man. They will tell you that there was an engine built by Miller but that the design came out of the head of a Milton, a Murphy and a lot of other members of the racing fraternity. As for the Duesenberg cars, there never were any brains behind them but Fred's.

Others watched the entire scene with impartial fascination. Dick Loynes said:

> There were two great clearing houses of ideas concerning racing technology in those days. Miller in the West and Duesenberg in the East. They were the targets, and practically the only ones, for all of the good and bad ideas of the entire racing fraternity. This was an extremely important part of their strength.

And ideas were wherever you found them. Said Eddie Miller:

> It was a big spy game. We all spied on each other. I painted false ratio numbers on my gears just for the opposition to read.

There *was* a Duesenberg cult as there never was a Miller cult. Fred was gregarious and warm, and just to be in Augie's presence was to feel affectionately toward him; they were natural father and brother figures. Miller, on the other hand, was reserved, remote, autocratic, and hard to get to know. Then, Duesenberg produced passenger cars that were legendary in their own time and were literally worshiped by tens of thousands of Americans. Miller toyed with the thought of producing as good or better road machines, but the required discipline and organization were totally beyond his inclination or capacity.

Fred never hesitated to help himself to Miller ideas as freely as Miller dipped into Duesenberg's store. With mere months left for the three-liter formula after Murphy's Indianapolis victory Fred launched an entirely new cylinder-head design for the Duesey 183. He abandoned the single camshaft and three valves per cylinder and adopted dual overhead camshafts, four valves per cylinder, cup-type cam followers, and spark plugs that were centrally located in the precise Miller manner. In other words, he adopted the entire Miller top end, while retaining his shaft-and-bevel camshaft drive.

Ralph de Palma was given the job of test-driving the prototype head in the same car with which he had lapped Indianapolis at 98 MPH in the last race. With no other change than the DOHC head he lapped the course at an easy 107 MPH. Tests showed

Model A passenger car and 122-cubic-inch race car were truly variations on each other's designs. JERRY GEBBY

that the central position of the spark plug—with its better cooling as well as its location relative to flame travel—permitted about 15 PSI higher compression before the plug was likely to be burned. The other advantages were more or less equally dramatic, and Duesenberg went almost the whole Miller way by converting to gear-train drive to his camshafts. He retained many traditional Duesenberg features, including the still-reliable three-bearing crankshaft, detachable cylinder head, and battery ignition. And he added such new features as a one-piece aluminum crankcase and block with steel liners. While learning from each other the arch-rivals were doing their utmost to out-think each other.

They entered the 1923 season and the two-liter, 122-cubic-inch formula with these closely similar designs, but Duesenberg was poorly prepared and by the end of the season still had not gotten the combination sorted out.

Indianapolis was a fiasco, even though the pace car was a Model A Duesenberg phaeton, driven by Fred. Sometime after midnight, with the race due to start in a few hours, one race car left the plant for the track. By special arrangement it was qualified just after daylight. At about that time two other race cars left the factory but were almost immediately blocked by the early morning traffic jam and never got near the Speedway. The car that did make it was new and tight and was driven to tenth place by Wade Morton and Phil Shafer at an average of 74.98 MPH. Even L. L. Corum's Fronty-Ford Barber-Warnock finished fifth with a speed of 82.58. Duesenberg won nothing in 1923; Miller won everything.

This time it was Miller who was directly indebted to Colonel E. J. Hall, who had suggested using just two valves per cylinder in truly hemispherical combustion chambers in place of the traditional four valves in pent-roof chambers. Duesenberg adopted this winning configuration of Miller's for 1924 and with it—and with his

Ditto. These excellent (though screened) shots are the only ones available. JERRY GEBBY

terribly hard-won mastery of the centrifugal supercharger (dealt with in detail in Chapter 22)—finally achieved his first Indianapolis victory with Corum and Boyer sharing the wheel and averaging 98.24 MPH for the distance—a beautiful new record. But except for that historic and prophetic performance, from which Miller profited almost instantly, the only other win for Duesey on the Championship Trail that year was Shafer's 70.1 MPH in the 150-miler on the Syracuse mile dirt track on September 15.

The 1925 season was the opposite: one of glorious revenge for Duesenberg and of brilliant achievement for Pete de Paolo and the car he called his "Immortal Banana Wagon."

He clocked 135.0 MPH on the Culver City boards on April 19. On the 30th at Fresno he won the annual 150-miler. At Indianapolis on May 30 he averaged 101.1 MPH—a new absolute record for the distance, the first to exceed the 100 MPH barrier for 500 miles. On June 13 at Altoona he won with a 115.9 average for the 250 miles. On July 11 on the Laurel, Maryland, board oval he won the 250-miler at 123.3. He topped off the season at Salem, New Hampshire, with 125.2 MPH for 250 miles. He won a magnificently earned AAA National Championship for himself and Duesenberg.

That year de Paolo took part in the Italian Grand Prix at Monza, being honored with a position on the then unbeatable Alfa Romeo team. But Alfa's most feared competition was from the Duesenberg entries of Pete Kries and Tommy Milton. Kreis was quickest in practice and turned the fastest lap of the 490-mile race at 101.5 MPH. But the Duesies' gearboxes had not been designed with constant shifting in mind and both failed, although Milton managed to finish fourth on top gear only.

A highly significant feature of this race was Europe's first exposure to the one-man cars that the Speedway Era had just produced. In its September 15, 1925, edition *Auto Italiana* observed:

> The Duesenbergs had a great advantage in their extremely narrow bodies, although this streamlining was spoiled by the modifications required by the absurd regulations. Other advantages of this type of car are its logical central steering and shifting. It is to be hoped that in the coming year we will adopt this system and that it will be adopted without irrational compromises.

It was not until the free formula of 1931 that *il tipo monoposto degli americani* could be adopted by European race-car builders.

The period of near total Miller domination in the United States began in 1926 and coincided with the change to the 1.5-liter, 91.5-cubic-inch formula, the summation of the whole of racing history up to that point. For it, Duesenberg approached Miller engine design even more closely. He dropped his traditional one-piece block-crankcase in favor of a cast-iron block on an alloy case. He also finally arrived at the conclusion that five main bearings were better than three.

These supercharged engines were no match for Miller's in 1926. De Paolo won the 300-mile opener of the season (still for 122-cubic-inch machines) on the short-lived Fulford-Miami boards at 129.3 MPH. The 91's made their bow at Indianapolis, and de Paolo drove the best-placed Duesenberg: he came in fifth, behind a sweep of Millers. The marque's only other honors for the year were Fred Winnai's win in a 25-miler on the Langhorne, Pennsylvania, dirt track plus a new mile dirt-track record which he set there at 90.2 MPH.

The next year was better. There were no victories until Memorial Day, but then rookie George Souders roared home first at Indianapolis at the wheel of the ex-de Paolo 122, now equipped with a 91 engine. Souders' average was 97.54 MPH, nearly four MPH faster than Earl Devore's second-place Miller.

Five Duesies started in this race but rookie Babe Stapp broke a universal joint

Joe Boyer, a Detroit millionaire, co-drove supercharged car to 1924 Indianapolis victory with L. L. Corum (upper right with Fred Duesenberg). IMS

on Lap 24, then relieved Ben Shoaf. Stapp, in Shoaf's Perfect Circle Duesenberg Special, had second place secured when, just five miles from the finish, his rear axle gears failed. The nearest pursuer was ten miles to the rear. The Shoaf machine was, along with Wade Morton's Thompson Valve Special, one of the new radically offset Duesey chassis with ring and pinion gears located adjacent to the left rear wheel. The gears had to be specially machined to accommodate the diagonal drive shaft, and they turned out to be the weak link in the system. Morton had been relieved by Fred Winnai when, on Lap 152 and perhaps also due to failure of the unusual rear axle gears, the car hit the wall and burned. Duesenberg's best showing in the "500" was Dave Evans' fifth place in a car with a conventional rear axle. The sixth place credited to de Paolo and Duesenberg at the time was in error; Pete's car was a Miller front-drive. However, he did give the marque its only other win of the season with an average of 116.6 MPH for 200 miles at Altoona on June 11.

In 1928 Duesenberg's luck was equal or worse. The cars of Slim Corum and Dutch Baumann were wrecked in practice at Indianapolis. Leon Duray qualified his Miller front-drive at 122.391, but the fastest Duesey was Jimmy Gleason's at only 111.708. In the race both Benny Shoaf and Ira Hall were doing well but both were eliminated by stupid minor collisions which were not their fault. Wrote Jerry Gebby in the *Auburn-Cord-Duesenberg Newsletter:*

> But the real catastrophe occurred when Jimmy Gleason stopped at his pit on Lap 195, only twelve and a half miles from the finish. He had held a nice lead in first place for the past forty laps, but the engine was dead when he coasted in. Steam was pouring from under the hood and the cause was found to be a water hose that had come off the manifold, allowing water to run onto the spark plugs and distributor. While the engine was being dried, re-filled and started again, the entire money field went past the unfortunate Gleason, plus two more cars. He left the pit in thirteenth place and was unable to make up any of his loss in the remaining five laps. Twenty thousand dollars went down the drain with that loose hose, but automobile racing has always been this way.

Fred Frame finished eighth in that race in a privately owned Duesey. The rest of the year was a washout except for Winnai's slow 101 MPH win in a 100-miler at Atlantic City.

In 1929 seven Duesies were entered in the "500." Bob Robinson's was not finished in time to attempt to qualify. Thane Houser's supercharger drive broke with no time left for repair. Frank Swigert's 99.585 MPH was too slow to get him in the race. Ernie Triplett had the fastest qualifying time for a Duesenberg at 114.789, which was none too promising in view of the many faster cars in the field, up to and including Cliff Woodbury's Miller at 120.599. The other Duesies to make the race were those of Gleason, Winnai, and Bill Spence.

The race got off to a bad start for all concerned, particularly for Duesenberg, when on Lap 9 Spence skidded into the wall at high speed in the southeast turn. The car overturned and slid, crushing the driver fatally. On Lap 47 Triplett left the fray with a broken connecting rod. But Gleason finished third, less than four MPH slower than the winning Miller's 97.585. His performance was particularly admirable since his car was a 1923 machine whose 122-cubic-inch engine had been replaced with a 91 which was equipped with one of the early superchargers. Winnai, in a newer car with one of the latest blowers, finished fifth. Elsewhere on the Championship Trail that season the best performance by the marque was Winnai's third in the Syracuse 100-miler.

In 1930 the old order changed and the 366-cubic-inch formula was introduced. Fred and Augie, neither of them with any great hopes for their racing future, decided to take different paths. Fred chose to do his racing with modified Duesenberg Model A passenger-car components, while Augie remained faithful to the wiry, small-displacement thoroughbreds.

Between them and several private owners nine Duesenberg-powered cars were qualified for the 1930 "500." Augie's cars for Dave Evans, Deacon Litz, and Babe Stapp had displacements of 138, 150, and 142.5 cubic inches respectively.

In Fred's camp William Denver and Rickliffe Decker showed up with two of the 1923 Indianapolis Mercedes, which were later owned by Louis Chevrolet. According to *Motor Age* these two cars were fitted with the 300-cubic-inch engines from Tommy Milton's Land Speed Record machine. If this was indeed true the three-valve-per-cylinder, 1919 engines did not qualify under the new rules and were replaced with slightly modified Model A Duesey engines. The Model A-powered machines were the fastest qualifiers in the semi-stock class, with Bill Cummings averaging 106.173 MPH for his ten miles. The quickest of Augie's enlarged 91's was qualified by Deacon Litz at 105.755. But no Duesey was even close to the eventual winner—Billy Arnold in a Hartz-Miller front-drive which qualified at 113.263. Still, anything can happen in a race, and Duesenberg chances were good. But everything happened to Duesenberg, almost from the moment that Wade Morton pulled his Cord pace car off the course at 80 MPH. *Motor* reported:

> On the very first lap, on the first turn, not half a mile from the starting line, Chet Gardner in a Duesenberg took a skid which might well have tangled up at least half the thirty-eight cars in what might have been the greatest catastrophe in automobile racing history. . . .
>
> As it was, a spill occurred early in the race which involved seven cars, while others evaded the mixup by a hair's breadth. Red Roberts, on his 20th lap, went into a spin entering the north turn (after leaving the back stretch). Peter de Paolo had relinquished the wheel of this car [Duesenberg] on the 8th lap because he said it was hard to steer. Stapp [Duesenberg] hit Roberts. Litz [Duesenberg] hit Trexler. Lou Moore climbed the wall. Johnny Seymour's left front hub momentarily locked into the wires on Jimmy Gleason's right rear wheel and later Seymour's car hit the wall. With wheel locked, Gleason stepped on the gas and as a result broke a timing gear and bent some valves.

This early elimination of three cars was only the beginning of disaster for the

In winning the 1925 "500" Pete de Paolo was the first driver ever to exceed 100 MPH for the distance. Supercharger expert Dr. Sanford Moss stands below flag. IMS

Duesies. On Lap 29 Cy Marshall in one of the Model A's spun out on the north turn and crashed through the retaining wall. His riding mechanic, his brother Charles, died on the spot, and Cy was gravely injured. Rick Decker, driving one of the Mercedes Model A's, skidded on Lap 42 and smacked the wall with his right rear wheel. He came into the pits, spent four minutes checking for possible damage, then took off again. But on the next lap the car's torque-tube yoke fractured and it, too, was eliminated. Almost simultaneously Bill Denver's sister car threw a connecting rod.

This left only Cummings' Model A and Evans' 138-cubic-inch "91" to defend the Duesey name. They did a beautiful job, Cummings finishing fifth in his rookie year with the fastest of the semi-stocks at 93.579 MPH. Evans came in immediately behind him with an average of 92.571 for the 500 miles. The only Duesenberg victories in the Big Time that year were Cummings' in the opening race of the season, a 100-miler at Langhorne, and in the final event at Syracuse, for the same distance. Both were one-mile dirt tracks.

Duesenberg's years of glory expired with the 91.5-cubic-inch formula. Fred Frame did finish a splended second at Indianapolis in 1931 in one of Augie's enlarged 91's—prepared, significantly, by wizard Harry Hartz. After that, only bigger engines of new design could hope to be competitive, and they were not forthcoming from the house of the Big D, the Winged Eight.

chapter 13

DUESENBERG: TOMMY MILTON AND THE LAND SPEED RECORD

EACH NEW ATTACK on the record for ultimate speed is the horizontal equivalent of an assault on an unclimbed Everest. Each new attempt is a voyage into the perilous unknown, and each one takes much of the courage that a man possesses. Camille Jenatzy's 65.79 MPH absolute record was just as much a death-defying feat in 1899 as Bob Summers' 409.695 with piston engines or Craig Breedlove's jet-propelled 600-plus in 1965. Again, the world always wonders, "Why do they do it?" and again the answer goes much deeper than the hope of material reward; again it is rooted in the passion and challenge of the machine and in proving one's self to one's self and to the world. All that Tommy Milton ever got for pushing the Land Speed Record to 156.046 MPH in 1920 was a sterling silver tea service from the Goodyear Tire & Rubber Company.

At Uniontown on September 1, 1919, the Great Milton (only one other driver, Earl Cooper, ever amassed more Championship points in his career) seemed to have the 225-mile race won hands down. He was a full lap ahead of the second-place Gaston Chevrolet—Joe Boyer Frontenac as he flew down the grandstand straightaway at well over 100 MPH. Then a great gasp went up from the crowd of 40,000 as a cloud of greasy black smoke erupted from beneath the Duesenberg's hood, and sheets of raw, red flame enveloped its cockpit. Milton reacted automatically. With one hand he

After the successful record runs on the Daytona sands. Bare spot on hood is a result of engine-compartment fire. BILL TUTHILL

warped the front wheels over to full lock, and with the other he yanked the hand brake and locked the rear wheels. The brake lever was a stout steel bar but it bent like a lead pipe under Milton's madly adrenalized grip. The car pivoted on the proverbial dime, rolled backwards as the wind blew the flames away from the driver and riding mechanic Dwight Kessler, and came to a stop. The gravely burned pair were pulled from the car and rushed to the local hospital. Milton refused to authorize the doctors to amputate his leg, spent nine weeks there on his back, then left on crutches for his great record-breaking rendezvous with the 16-valve aero engine at Sheepshead Bay. Before that a little history was made in Milton's hospital room.

Jimmy Murphy came to visit him, in fact, to say good-by. Murphy, who happened to be an orphan, was a very complex young man: wistful, gentle, disarming, totally charming, and, beneath all this, aggressive as hell. Milton liked and befriended him from the day the twenty-three-year-old Murphy took a mechanic's job with Duesenberg early in 1919. "It takes one to know one," and Milton recognized Murphy's class, made him his riding mechanic, and agreed that he had the stuff of which good drivers are made.

At the Uniontown Hospital Murphy unburdened himself to his mentor. It had become obvious that Fred had no intention of granting his ambition, so Jimmy was going back to his old garage-mechanic job in Vernon, California. Milton urged him to hang on until the start of the 1920 season. In the meantime he, Milton, the captain of the Duesenberg team, would lay down the law to Fred that either Milton's protégé was given a car to drive or he, Milton, would quit. Murphy stayed. Milton was good to his promise. Murphy got his ride and more than vindicated Milton's faith in his ability.

While Milton lay on his back for better than two months he had nothing but time in which to plan and dream, and it was then that he hatched his plan to become the fastest man on earth. The 300-cubic-inch formula was dead, and Fred had three big straight-eight engines which now were useless. Assuming correctly that he could make use of them for the good of the cause, Milton whiled away the long days designing a 16-cylinder, 600-cubic-inch record machine. When he communicated the plan to Fred, along with his willingness to shoulder its cost, he was told to fire up his crutches and hustle to Elizabeth.

I had many wonderful sessions with Tommy Milton, and we corresponded for years. At one point I praised his unique and vivid style and he replied:

> I am a bit unhappy to learn that I wear my ego so conspicuously. As a matter of fact I do like language and, who knows, but one day I shall be so brash as to take a hand at an auto-biography. Inasmuch as I have been thinking about this for some twenty-five years it seems a reasonable presumption that it will never occur.

This is the reason for the liberal use of long quotes in this book. Nothing is easier than to paraphrase source material. It makes the author seem authoritative but at the cost of true, first-hand authority. There is no substitute for authentic experience nor for the vitality of language such as Milton's.

> In the construction of this car I was one of the busiest little beavers you ever heard of. I sawed out the frame from flat stock and swung a 16-pound sledge for the blacksmith who formed the rails. The power plant consisted of two straight-eight competition engines which were connected directly to a common solid rear axle which had two pinions and two ring gears. Fortunately, we were then using cone clutches. I say fortunately because the Contest Board had told us that no reverse gear would be required. When we had arrived at Daytona there was a reversal of this opinion which I need hardly tell you posed a pretty serious problem. We hurdled this by attaching a flywheel starter gear to the cone of one of the clutches and by making a leather wheel which, with benefit of Rube Goldberg linkage, we were able to press into the crotch of the two flywheels. Since the arrangement was the plan of Fred and Augie Duesenberg, I need feel no embarrassment in stating that it was by any yardstick an ingenious solution. The thing actually worked and movies were taken of the car going backward on the beach.
>
> The solid rear axle was the first one I had ever seen that was live. In other words, the axle shafts themselves supported the weight and were not contained

in the usual conventional housing. The bearings were carried in the outer ends of the differential casings. Many competition cars have subsequently been built with this type of axle, but this may have been the first one.

Evidently the 300-cubic-inch engines had been stored in Los Angeles because when the team went to contest the Beverly Hills 250-miler (Murphy's first victory) on February 29, 1920, the beach-car chassis was taken along, and the engines were installed there.

> Due to my Uniontown crash and the investment in the beach car my funds were depleted and so I welcomed the opportunity to pick up a few dollars racing at Tropical Park in Havana. I sent Murphy and Harry Hartz with the beach car to Daytona to get it ready for the record attempt.
>
> When I got back to Key West and picked up the morning paper it said in screaming headlines that Murphy had smashed de Palma's world record of 149.9 MPH with a speed of 152—in my car, which I had paid for, had actually built and was paying him to be there preparing. I don't think that the world ever has looked so black before or since. I could have killed him when I got there; he understood that and got out.

Milton slaved over the car and ran and ran but could not get up to Murphy's speed. The gossip grew that Tom didn't have the guts to get his foot into the carburetor as far as Jimmy had. So he finally decided that his only hope was to tear the engine down to the last lock-washer, rid them of the sand that they had inhaled, correct all their clearances, and make a final try. The car's shelter was an open shed on the beach through which sand blew continuously. Milton rented enough tarpaulins to erect a tight enclosure, and he and Hartz did the job. If this effort didn't help there was nothing to do but pack up and go home in defeat. During the rebuild Milton got a steel sliver in his one good eye, which hardly helped. Finally the car was as ready as it ever would be.

The men of Duesenberg never had enough money to do things in a first-class manner and the crew at Daytona was a mere handful. The town offered no aid in the policing of the course and the various approaches had to be guarded by whatever voluntary manpower could be scrounged. Milton waited at the south end of the course until he felt that all hands had had ample time to reach their guard-posts on the approach roads to the beach, and then he charged off on his first official run. As he headed, full-bore, under a pier that crossed the course, a citizen in a Model T Ford touring car pulled up precisely in the path that Milton was committed to follow. Seeing the projectile hurtling toward him, the wandering motorist seemed to panic, and Milton watched him futilely jumping up and down and waving his arms. Milton made the slightest of swerves, which left the Model T half-buried in sand, and kept his foot all the way down until he crossed the finish line. Back to Milton:

> The car had its faults. One was that the exhaust pipe from the left-hand engine ran through the cockpit. The heat was pretty terrific. Then, the centrally mounted steering column was an open tube. One of the engines had its crankcase

breathers in the center and when I got under way oil vapors came streaming up through the hollow tube, covered the windscreen and my goggles, with the result that visibility became practically zero, zero. Actually I finished that northbound run by using the spectators on the beach as a landmark.

The return run went well enough until smoke, heat and fumes announced that we were on fire; I was carrying a full-length underpan in which there was undoubtedly oil and gasoline to support the blaze that cooked the paint off the hood. I took a squint at the aluminum firewall, which seemed not to be melting. The zero station was not too distant and it was, of course, highly desirable to finish the mile since I had no way of knowing how serious the fire damage might

Right: Milton with record machine in exhibition appearance in Indianapolis. Note how upper exhaust pipe passes through cockpit. IMS

Below: Power source of the LSR Duesenberg was a pair of straight-eight, 300-cubic-inch SOHC racing engines. TOMMY MILTON

Note hollow steering column, path of exhaust pipe from left-hand engine, reverse-gear device between and above flywheels, direct drive to two final-drive units. TOMMY MILTON

be. Having finished the mile I got close to the water's edge with the idea of dunking her but this did not prove necessary. It was fortunate that we were able to complete the run because there was considerable damage—probably two or three days' work—and the season when good sand can be expected already had passed. Extinguishing the blaze was laborious because the hood was bolted on; when finally removed the fire was put out with beach sand. Having been rather severely burned at Uniontown just a few months before, this experience was certainly unwelcome, if not alarming. But it did give me the satisfaction of beating both de Palma *and* Murphy.

Although this resulted in practically a total breach between Murphy and myself, the mellowing influence of time impels me to say that there were extenuating circumstances. Those who knew Jimmy will attest to the remarkable personality which was his, and which influenced Fred Duesenberg in particular. I know beyond any peradventure of doubt that Fred urged Jimmy to run the car all-out. Furthermore it is—or at least it was then—the fervent ambition of any youngster to become the World's Speed King. Undoubtedly the temptation was tremendous to the extent that, momentarily at least, it overrode his evaluation of loyalty. At least in retrospect it was purely a personal feud and I have always been happy that the conflict did not get a public airing. The final word on the matter is that after his untimely death at Syracuse in 1924 I was privileged to escort the body to Los Angeles.

Milton was modest. Actually it was he who came forward and shouldered all the responsibilities attendant upon Murphy's death. As for Fred Duesenberg, he was finished with the man. He was committed to drive at Indianapolis a month after Daytona. He finished an excellent third, after Gaston Chevrolet's Monroe and René Thomas' Ballot, then resigned from the team.

chapter 14

DUESENBERG: WHEN AMERICA WON THE FRENCH GRAND-PRIX

FOLLOWING THE THOMAS FLYER'S WIN in the strange New York to Paris "race" of 1908, via North America and Asia, American contestants wisely stayed on their own side of the water. After the first World War, however, it quickly became clear that the quality of American racing machinery had been radically transformed.

Some powers in France made it known that American cars would be warmly welcomed as entries for the 1921 Grand Prix of France, Europe's and the world's most important race. But others pushed through regulations that were designed to make this a one-make race, meaning no race at all. Instead of preliminary time trials for the weeding out of final entries, dynamometer tests were decreed. It was bad enough that hardly anyone had a dyno, least of all the Automobile Club of France, but it also was required that the three-liter engines should develop no less than 30 BHP at 1,000 RPM and 90 at 3,000 RPM. This, in effect would give Ballot the victory without his ever having to bring his cars to the starting line. Said *Automotive Industries:*

> Owing to the general dissatisfaction with the regulations the outlook for the 1921 Grand Prix is anything but bright. There is very little chance for a team from this country competing, because none of our large manufacturers participates even in our big classic, and to most of those who have been building racing cars

Starting lineup of the 1921 French Grand Prix. Murphy is Number 12. GB COLLECTION

in America during the past several years the expense of a Grand Prix venture would undoubtedly be prohibitive.

This was a very chaotic period, made more so by the fact that the Automobile Club of France labored under the impression that racing in the United States was controlled by the small New York organization which was called the Automobile Club of America. Finally, with the assistance of the ACA, the ACF was persuaded that the Contest Board of the AAA was the national governing body, and meaningful dialogues could begin. The end result was that the ACF, through Charles Faroux, advised the AAA that it had adopted the current Indianapolis rules in their entirety and, to make things more attractive, planned to stage its Grand Prix in a part of France where American popularity was at its highest. The bench-test requirement was abandoned and the tricolor carpet rolled out. Still there were no Stateside takers.

The situation was nearly as grim in France. Ballot had a strong four-car team and a magnificent stable of drivers: Louis Wagner, Jean Chassagne, Jules Goux, and Ralph de Palma. Fiat announced that it would enter but in the same breath let insiders know that it had no intention of entering. Sunbeam-Talbot-Darracq announced its participation with no less than eight cars. But then chief engineer Louis Coatalen chose this time to go off to relax on the Isle of Capri, the various shop heads fell to

Olson, Augie, and Murphy, setting sail for France; Murphy's hand had been burned in a minor accident. GB COLLECTION

engaging in knock-down, drag-out physical combat, and the cars did not get built. At the eleventh hour two British office employees took the situation into their own hands, whipped the troops into line, worked with the tools themselves night and day, and got some cars ready. Then there was a single 1.5-liter Mathis entered for the ride, and that was all.

Of course American passenger-car manufacturers had nothing that they could pit against even this modest field. But Fred Duesenberg thought that he might have something, and it tied in with his own new passenger-car program. One fine day he shot a cable to W. F. Bradley: "ENTER THREE DUESENBERGS IN GP." He neglected to send entry money.

Bradley knew full well that the Duesenberg name was one to be reckoned with. He went directly to the ACF's palace in the Place de la Concorde, where it still is today, and encountered Baron René de Knyff in front of its massive portals.

"I have three good American cars for your race," he said. "The entry fee hasn't come through yet, but if you'll give me just a few days I know you'll have it."

De Knyff, a compulsive stickler for form, cast his eyes up at the palace and said, "Why don't you just ask me to set fire to this building?"

Bradley was doing the man a huge favor, bringing him cars to fill out a race that very likely never would start. He felt that he was entitled to this small courtesy in return, but it was haughtily refused.

The deadline was at hand, and Bradley cabled Duesenberg to this effect, advising him that there still was a month's period of grace during which entries could be made, but at double the fat entry fee. The man who paid it and for a fourth entry to boot and for the transportation of the team and the cars was none other than Albert Champion. His was a Franco-American rags-to-riches success story, and it pleased him to spend a little of his American-made wealth on this cultural mission.

Meanwhile, due to their formidable internal difficulties, the STD team was withdrawn, as Bradley had known it might be, and Fiat bowed out officially. This left only the Ballots and the lone little Mathis. The ACF prepared to cancel the race the instant the grace period expired.

Bradley could be as difficult as those he had to deal with, and he held the Duesenberg entry money for a solid week. Came the final day, with the French Grand Prix promising to be a total fiasco, and Bradley still stayed away from the Place de la Concorde. De Knyff's office closed at six P.M. and Bradley waited until the first clock began to toll. The bell towers of Paris all were minutes apart and he knew what he was doing when he walked past the ancient little bird-cage elevator that still goes to the CSI offices on the top floor and headed up the stairs. The last clock was striking when he flung open the door, tossed the money on the desk, and said to the Baron, "This is for four Duesenbergs. Now perhaps you'll have a race."

That saved the first postwar French Grand Prix. Then, twenty-four hours before the start, STD announced that it would be competing after all, with four cars.

Next door to the ACF is the majestic Hotel Crillon, the old palace in which the Treaty of 1778 was signed, by which France recognized the independence of the United States. In 1921 it was the most distinguished hotel in Paris and the haunt of the entire diplomatic corps. It was there that Ralph de Palma told Ernest Ballot he wanted to stay, and his patron approved. De Palma, who still wore the aura of one of the greatest drivers of all time, was as much a natural aristocrat as he was a born showman. He was one of the most cherished guests ever hosted by the Crillon, which was in a position to choose its guests and did so. In Bradley's opinion the United States never had a more effective ambassador.

Duesenberg, as usual, arrived in cliff-hanger style, with only hours to spare but, for once, with excellent advance preparation. American race driver George Robertson, a giant of the old Vanderbilt Cup days, had spent the war years in France. He held the rank of lieutenant colonel with the Bolling Mission and spent most of the duration managing the movement of American military vehicles in and out of the important port of Bordeaux. Fred Duesenberg had secured Robertson's services as team manager

The winner at speed. ERNIE OLSON

in France, and he sent Augie over to take care of the mechanical side of the campaign as he alone could. The press agreed that while the team suffered from lack of road-racing experience its organization was far superior to that of its rivals.

The Duesenberg drivers were to be Jimmy Murphy, Joe Boyer, Albert Guyot, and Louis Inghibert. The course was laid out over 10.7 miles of public roads, consisting of stone beds over which a sand composition had been spread and rolled. It was a slippery surface, and practice was filled with incidents.

One of the Ballots crashed, was a total loss, and its driver, Renard, was killed. This entry was replaced by the only other available Ballot—a two-liter four-cylinder, which Goux drove. Murphy's brakes locked when entering a turn, his car got sideways and landed upside down in a ditch; putting him in the hospital with a broken rib. Team driver Inghibert had been riding with Murphy, and his injuries were so serious that he could not hope to drive in the race. His Duesenberg was turned over to wine magnate André Dubonnet, all his life an avid motorist, who, incidentally, gave General Motors its first independent front-suspension system and later became one of the directors of Simca.

During the days of practice French carburetor manufacturer Claudel made it financially interesting for all entries to use its products. Murphy alone among the Duesey drivers refused, preferring to stick with his tried-and-true Miller carburetor. This proved to be a wise move because during the race both Duesenbergs and Ballots had trouble with grit jamming their Claudels' rotary throttle valves. For all except Murphy time had to be taken out during the race for the fitting of supplementary throttle-return springs.

Important changes had taken place in tire technology. The straight-side tire had come into being, in opposition to the old clincher type, but it was as controversial and unproved as the balloon tire would be until Memorial Day, 1925. Ballot had exclusive use of Pirelli's new straight-sides and Duesenberg came loaded with gum-dipped Oldfields, products of Barney's association with Firestone. The STD's used conventional racing Dunlops. All seemed equally good in practice.

It has often been said that Duesenberg's great advantage in this race lay in its four-wheel brakes, but this is incorrect; *all* entries had them. The French used the Perrot system of ball-joint mechanical linkage, which Bendix was to adopt in the United States.

What Duesenberg did have that was unique was the Wagner-type hydraulic system, taken from the Model A passenger-car prototype. It was a novelty that attracted almost no pre-race attention, but its advantage of perfectly equal effort applied to each of the four brakes and therefore much more efficient braking and automatic adjustment permitted the Duesies to drive significantly deeper into the course's many turns. But the system still was full of bugs and only was made to work properly during the practice period. Ernie Olson recalled:

> At first we had too much braking on the front and every time Jimmy would hit the pedal the car would start fishtailing all over the road. Finally he landed in

the ditch and got badly banged up. We all worried about this crippling problem. As I recall we had 14-inch drums front and rear. I noted that the Ballots' four-wheel brakes were working very well and that their front drums were smaller than the rear. I went to Augie and suggested that we try the same effect by getting rid of some of our front brake lining. He told me to try it and I took a hacksaw and chopped two inches of that metallic lining from each of the front shoes. Jimmy was still in the hospital and I got Joe Boyer to take the car out and try it. We hurtled down the straight, he hit the brakes and the car just squatted. It tried to bury itself in the road without a bobble. We really had it over the French after that.

The Americans learned something else in practice. The French all carried spare tires and wheels. The Americans ran tests that proved to them that it was quicker to drive on the rim, even if a blowout occurred just after the pits, than to change wheels on the course and then stop again to pick up another spare wheel and tire. The French would not consider running without spares; it had never been done.

Among all the events of the practice days perhaps the most decisive was the final rift that occurred between Ralph de Palma and Ernest Ballot, although Ballot himself never was fully aware of it. De Palma was recognized universally as one of the world's finest and fastest drivers, which is why Ballot had hired the foreigner. It was not that Ralph was too lazy to change gears for himself. It was simply that he, too, took a highly analytical approach to his craft and had proved to his own satisfaction that he could gain perhaps a tenth of a second on every gear change if he would keep both hands on the wheel, nod to his riding mechanic, punch the clutch pedal, and let the mechanic do the shifting. Those tenths could add up and win races. De Palma was anticipating the preselector transmissions of the Thirties, and the automatic transmission revolution brought about by Chaparral in 1965.

Ralph knew the imperious Ballot well enough to know that he never would authorize such unorthodoxy. So Ralph explained to him that he shifted best with the lever on top of the gearbox, American style, rather than against the frame in the standard European location. Ballot accepted that, the change was made, and Ralph and his nephew, Pete de Paolo, went off to perfect their technique. Nod, depress clutch, Pete throw the shift, release clutch, keep charging, gain those tenths. They quickly reduced it to a science.

And then one day during practice, bolting down the straight into the Mulsanne bend—a 90-degree that required downshifting to second and perhaps to first, Pete missed the shift. The gears grated and the engine over-revved wildly. And Ernest Ballot happened to be standing right there, taking it all in. He reacted with pure fury.

Ballot had the shift lever returned to its original position. He did not care that de Palma perhaps had a point; he would not tolerate discussion. And that was the end of the romance between de Palma and Ballot.

The car was the same one which Ralph had driven at Indianapolis less than a month before. It had a fresh engine, which it sorely needed, but Ralph swore that it was five MPH slower than the old one. He drove the race, but Ballot had shown contempt for his efforts at winning strategy; and therefore why should Ralph extend himself for a victory that his patron wanted only on his own know-it-all, haughty diploma-engineer terms? Ralph's heart was no longer in the race—there was no money in it anyway—and afterwards he told friends, "Well, it cost me some glory but at least I've learned everything about Ballot and his cars that anyone needs to know."

That knowledge, fed into Packard's last racing effort—for the coming 122-cubic-inch formula—was not enough. The beautiful little Packard never raced successfully, and de Palma's career in the big time was at its end.

The historic 1921 French GP took place on July 25 with Murphy, taped from hips to armpits, ready on the grid with Olson beside him. The thirteen starters were lined up in pairs, the little Mathis sitting alone in the rear. Each pair was flagged off at thirty-second intervals, and the pace was terrific from the start. De Palma and Boyer both averaged 78 MPH for the first of the thirty laps, followed by Murphy and Chassagne, neck-and-neck. The second time around Murphy and Boyer were in the lead and running marvelously. On Lap 3 Chassagne overtook de Palma, and the leaders remained in that order until the end of Lap 6 when Boyer decided he had better stop to check his tires.

The STD cars already were in severe trouble. The surface of the course began to disintegrate by the end of the first hour; the sand flung aside, leaving only the coarse rock of the roadbed. It was like driving in a hail of shrapnel, rocks the size of fists flying at closing velocities of far over 100 MPH. The STD tires disintegrated almost as fast as they could be changed. Their complete unsuitability was recognized immediately, and the team rushed through the pits and parking areas, buying, borrowing, begging all the tires they could find that would fit their rims. The problem was so severe that they quickly ran out of ready-mounted spares, and the STD cars had to wait idly in the pits while new tires were pried onto their wheels. These machines, though the heaviest in the field, were extremely powerful and fast, and with more rugged rubber their performance would have made for a very different story.

On Lap 7 Murphy also dashed into the pits to check his tires. Chassagne, on the Pirellis that proved to wear like iron, surged into the lead as Boyer roared back into second place. Murphy's stop lasted mere seconds, and he quickly overtook Boyer and went after the leading Ballot. But Chassagne remained in front until Lap 17, when he was forced to retire with gasoline pouring from a stone-riddled tank.

Murphy took over the lead, followed closely by Boyer, whose Duesey threw a connecting rod on Lap 18. Then Guyot's Duesenberg moved into second, a position which he might well have held for the remaining laps. Then a great, hurtling rock cracked his riding mechanic on the skull, and when he stopped at his pit the man was too dazed to be able to crank the engine. Old veteran Arthur Duray, just a spectator in the crowd, saw what was happening, vaulted the fence, pushed the mechanic aside,

Murphy about to lap de Palma's Ballot. ERNIE OLSON

Victory! Note stone-strewn surface. ERNIE OLSON

spun the engine into life, and jumped aboard alonside Guyot. But the delay had put him back into sixth place, and there he stayed.

Oil tanks, fuel tanks, and radiators were punctured by the raining rocks. De Palma's gas tank began streaming, but he clung to his second place and refused to stop. Then on Lap 29, while he was running far in the lead, a rock sailed through Murphy's radiator core. Almost simultaneously, a tire blew. There were twelve miles yet to go.

Murphy did not reduce speed, on the theory that centrifugal force would help to keep the tire off the rim. Every bolt and rivet in the car threatened to snap before he reached his pit. Olson leaped out to change the wheel, and Murphy kept the engine revving as water was ladled gingerly into the red-hot block. Olson took a swing at the wing-nut, missed, his hammer flew twenty feet, he retrieved it, got the wheel changed. Off they took again, with the water they had just taken on pouring from the ruptured radiator. Then, with eight miles yet to go, another tire went flat. But the white Duesey with the American flag on its side was 14m59s ahead of de Palma when it bumped across the finish line. Then, with a bone-dry engine Murphy managed to nurse his car for the required safety lap at about 40 MPH. Dubonnet's Duesey finished fourth and Guyot's sixth. The news flashed around the world, one report saying:

> American prestige had been enormously enhanced by Duesenberg's victory in France. The win was so fair, there was such an absence of the element of chance [!], that all open-minded spectators admit the best car won.

The French were not unanimously of that opinion, and *The Star Spangled Banner* was not played. Ernest Ballot was outraged. His cars were the best, by God, and he stood in the public square at Le Mans and harangued a crowd on his "moral victory." Someone shoved a wooden box behind him and he got up on it, proclaiming that the American cars had been reduced to junk while his cars were ready to start another thirty laps of the circuit. "Let's do it again!" he ranted, "and we'll see who wins!"

Guyot was in the throng, and when Ballot paused for approval he spoke out, calmly but loudly, "Engineer Ballot may be right. But I know only one winner in any race and that's the man who gets in first."

Ballot purpled, the crowd roared, and Ballot left his box and melted away. But the same sour-grapes spirit prevailed at the victory banquet, which Ballot dominated and where he again claimed a moral victory. Murphy was handed a small medal for his pains, and the Americans, some of them not too impressed with Old World sportsmanship and down to their last nickel again, headed home.

The race proved overnight the superiority of the straight-side tire, of four-wheel brakes, and of hydraulic brake operation. Three days after the event the French Rolland-Pilain Co. made a "descriptive seizure" of Murphy's car. The reason was "to protect the firm's legal rights in the system of braking." Rolland-Pilain had exhibited a form of hydraulic brake actuation at the Paris Salon in 1910, never had done anything with it but stated, not surprisingly, that it now planned to place its

The winner. ERNIE OLSON

system on the market. That farce was quickly laughed out of existence.

For the historic record and on the authority of Ed Winfield, the first known appearance of hydraulic brakes on an American racing car was on a 300-cubic-inch, four-cylinder Miller in the 1919 Santa Monica Road Race. The brakes failed due to leakage of the hydraulic fluid. The system was Lockheed's and already had been used experimentally on passenger cars.

W. F. Bradley wrote in *The Automobile* (New York):

> The recent Grand Prix race marked a revival of long-distance road racing in France, after the disastrous race in July, 1914, when Germany with its [4.5 liter] six-cylinder aviation engines won first, second and third. Murphy's victory averaging 78.1 MPH for 321.78 miles or 15 minutes faster than the Ballot, which finished second, left no ground for disputing that the best car won. This speed, compared with 65.5 MPH in 1914, indicates that advancement has been made during the past seven years, notwithstanding the fact that the German . . . engine . . . was well ahead of its time in engineering development.

A British report to *Automotive Industries* stated:

> The press-men have not failed to note that the Duesenbergs are production, or commercially possible cars and not special cars impossible to build at the price the public will pay.
>
> The result of this victory effectively disposes of the delusion that American cars are inefficient from lack of European experience in design, etc. It now only remains for the Duesenberg Co. to disclose its European price for this car [the Model A].

Within weeks after the race Model A Duesenberg orders were pouring into the offices of the firm's Paris agent. Had Fred's blossoming new enterprise had the ability to meet the demand at that psychological moment his personal story and that of the marque might have taken a very different turn. Instead, Eddie Miller recalled:

> The shysters were taking over everything. By early 1922 one of them had even taken over the race cars. If you wanted to race a Duesey you had to lease it from someone you didn't even know. There were no more parts, no support, and a lot of us said To Hell With It. But Fred and Augie kept fighting.

Albert Champion was very content with the lesson he had taught his former countrymen. Then one day as he was preparing to recross the Atlantic, a pair of gendarmes appeared at his door.

"Monsieur Champion?"

"Yes."

"You are a Frenchman." It was not a question.

"No, I am an American."

"But you were born in France and have not done your military service. You are under arrest as a deserter."

It was all the United States Embassy could do to get the spark-plug tycoon sprung from this rap. It was his last visit to the land of his birth.

chapter 15

DUESENBERG: TECHNICAL NOTES

AS WE HAVE SEEN, Duesenberg straight-eight practice goes back to the Patrol Model Marine Engine which was manufactured under license by the Loew Victory Engine Company, starting in 1915. The engine was used in submarine chasers and other fast craft and was essentially similar to the walking-beam four.

Then, late in 1916, Fred began dreaming of producing a luxurious, high-performance passenger car, and the layouts were made for a straight-eight automotive engine. Crankcase, block, and cylinder head were to be a single iron casting. The walking-beam principle was retained, and a three-main-bearing crankshaft and aluminum sump were adopted. This engine was very similar to the four, but the conventional split crankcase, rather than the barrel type, rendered it more suitable to volume production and easy maintenance.

The V16 aircraft engine of 1918 was a radical departure from all previous Duesenberg practice. It reflected the engineering talents of van Ranst, Bill Beckman, and George Dennis, with which Fred had strengthened his team. It also reflected the rich and varied experience with all manner of aircraft engines which the little organization encountered in the course of its government work. But it still retained valves at 90 degrees to the cylinder axis and, after problems with the V12, marked a return to the barrel-type crankcase.

In this engine for the first time Fred used three valves per cylinder. Instead of using two intake valves as Bugatti had, for example, Fred used just one at the top of

Engine of Murphy's 1921 French GP car: 183-cubic-inch shaft-and-bevel-driven SOHC, three valves per cylinder. IMS

his traditional combustion chamber. The two exhaust valves were placed below it, where they could be cooled by the incoming charge.

Each of the three valves per cylinder was operated by its own rocker, now too short properly to be called a walking beam. To achieve this obvious improvement the camshaft was moved very high in the valley between the two banks of cylinders. It was carried in its own very husky aluminum housing which, when bolted to the crankcase, added to the latter's longitudinal stiffness. The camshaft was driven by vertical shaft-and-bevel gears.

The V16's tubular connecting rods were of the fork-and-blade type. The five main-bearing crankshaft was of four-four configuration, as though two single-plane four-cylinder crankshafts had been joined, with their planes at 90 degrees to each other. Other configurations were possible, but this was the simplest and it worked very well. Bugatti had used it, but so had his predecessors.

This engine was as big as a small house. Four prototypes were built, two for the Army and two for the Navy. Two had direct-driven propellers and peaked at 1,800 RPM, and two were reduction-geared to turn the prop at 1,250 RPM. When Fred began scouring the industry for gears to hold such power—700 to 800 BHP, depending upon stage of tune—he found that the problem had never come up before. He finally solved it with the Maag system of gearing, used cogs with teeth four inches wide, and they never gave trouble. He pioneered the use of Maag gears in his Model A passenger car.

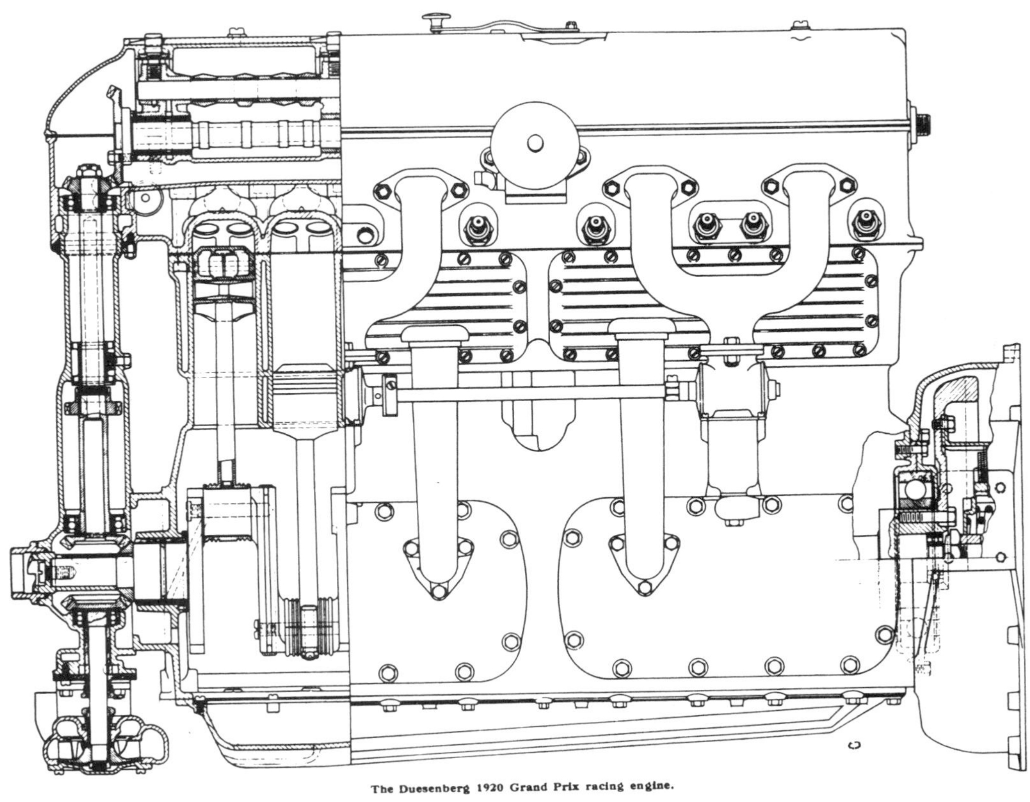

Same engine in longitudinal cross-section. KARL LUDVIGSEN

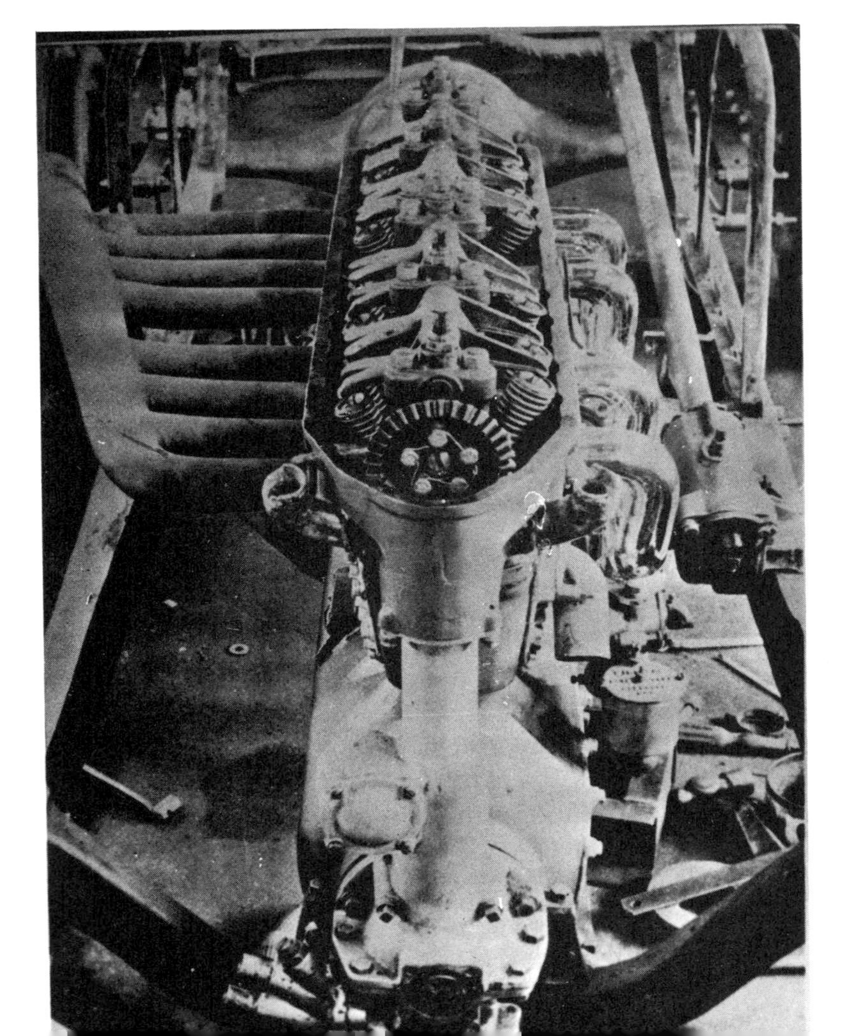

Same engine, showing details of valve train. JERRY GEBBY

Fred's walking-beam eight for this car had been designed during the war and, in the light of the V16 experience, its camshaft was moved very high in the block so that short, light rockers, 6.5 inches long, could be used. November 11, 1918, and the Armistice had scarcely arrived when patterns materialized, then castings, and three of these engines went into production at Elizabeth. But the engine already was obsolete.

No doubt remained concerning the superior efficiency of inclined overhead valves operated by one or two overhead camshafts, and so Fred rushed into the creation of a racing straight eight along these lines. Both Mercedes and Frontenac had demonstrated that one overhead camshaft could be competitive with two, so this was the choice that frugal Fred naturally made.

As in the V16 he used two exhaust valves and one intake valve per cylinder. Each pair of exhaust valves was operated by means of a single cam and a Y-shaped rocker arm. This measure in the interest of simplicity and economy was not as radical as it may sound, having been used successfully enough by Sizaire-Naudin, Delage, and Mercedes, among others. The Duesenberg forked rocker, however, did prove to be a weak point in the design. This was aggravated by the fact that the only proper method of adjusting valve lash on all the Duesey SOHC racing engines was the laborious one of filing the tips of the valve stems. To avoid this it was normal practice for mechanics to "adjust" these clearances by banging on the rockers with a hammer, which enhanced neither performance nor reliability. The camshaft was driven by vertical shaft-and-bevel gears at the front of the engine.

The barrel-type crankcase and the cylinders were a single iron casting with very large openings on the left-hand side and bottom of the crankcase for easy access to pistons and connecting rods. The cast-iron cylinder head was detachable in the interest of facilitating the foundry process, machining, and maintenance. Working with low pressures of compression and combustion, gasket failure was not a problem.

As he had done in most of his walking-beam engines, Duesenberg continued to use two spark plugs per cylinder, placing them on opposite sides of each combustion chamber. The four forward cylinders and the four aft ones each had their separate intake manifolds and single carburetor—a rotary-valve Miller. As already noted, the Duesey 300 used a two-piece, three-bearing crankshaft, with a bushing for the front main and ball races for the center and rear. The huge cone clutch of the four-cylinder period was retained.

After years of using magneto ignition Duesenberg switched to battery ignition for the eight. This was a highly compact aircraft-type unit, a special version of which was developed by Delco for automotive racing and which performed badly in the 1919 "500." The connecting rods, which always had been H-section forgings in the fours, were replaced by hand-forged tubular rods as had been used in the V12 and V16 aero engines. Most steel parts were made of chrome-nickel, this being the strongest alloy available at this period. Cobalt-chrome was used for the exhaust valves and low-percentage tungsten for the intake valves.

The 183-cubic-inch Duesenberg engine of the 1920 through 1922 formula was a scaled-down and much refined version of the original straight eight 300. Much study had gone into overcoming lower-end defects, and the entire lubrication system was greatly improved. The crankshaft now was made in one piece, with plain bearings front and center but still with a ball race at the rear. The crank webs between the diametrically opposite rod journals were machined from their billet in disc form and provided with integal counterweights. A groove was machined in the periphery of each web, and it communicated with the oil passages to the adjacent journals. A copper band (later steel) was shrunk over this groove and soldered in place in order to make centrifugal force help the pressure system feed adequate oil to all crank throws in the Peugeot manner. A trough was provided below the camshaft which provided an oil bath for the cams. The World War II Allison V12 engine, which used a top-end layout which was very similar to that of the SOHC Duesies, lacked this detail and suffered for it.

The four-four crankshaft, in spite of its inherent unbalanced couple, which tended to rock the engine about its central axis, caused no trouble for Duesenberg, nor for Miller, who copied it. Instead, it offered definite breathing advantages when the original two-carburetor system was used. But when this was replaced by a carburetor throat for each inlet port, and then by supercharging, the same crank configuration was retained along with its rocking couple.

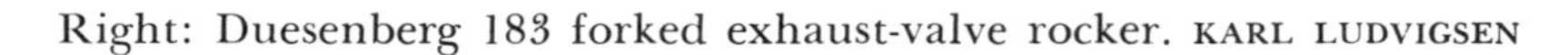

Right: Duesenberg 183 forked exhaust-valve rocker. KARL LUDVIGSEN

Below: Details of 183-camshaft design. KARL LUDVIGSEN

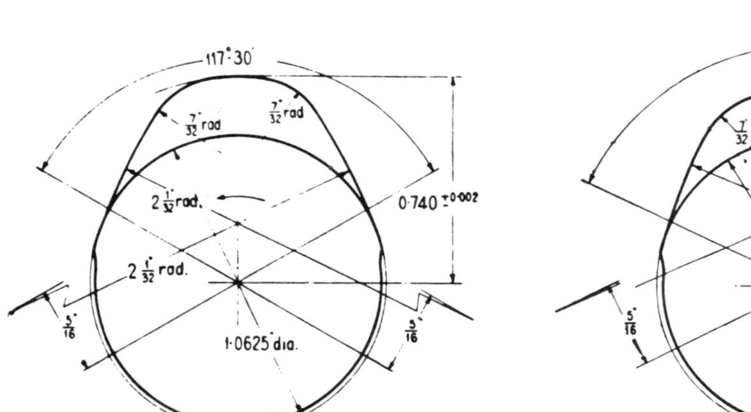

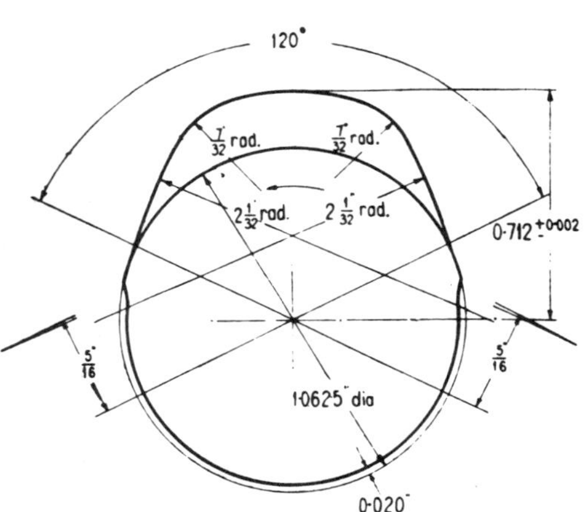

The pistons were of the semi-slipper type—not to reduce friction, however, but to permit removal of the rod-and-piston assemblies through the now-smaller hand-holes in the side of the crankcase and to allow the piston skirts to clear the circular crankshaft webs. As with some the walking-beam engines only two piston rings were used and the wrist pins still were locked in place by means of a bolt in the split small end of the connecting rods.

The vertical drive shaft for pumps and camshaft was simple in principle but was rendered intricate in order to compensate for variations due to thermal expansion and, most probably, for Duesenberg's notoriously variable machining tolerances. Mesh of the lower bevel gear with the crankshaft was adjusted with real difficulty by means of spacer washers of assorted thicknesses. The same shimming procedure was used for adjustment of the upper gears. The gears were integral parts of the shaft, and getting them to run true after hardening also was a problem.

Two exhaust valves were provided, as before, in the interest of improved cooling. Their throat diameter was .9375 inch and head diameter was 1.0625; lift was .3125. Intake throat diameter was 1.375, and lift was .375.

Dual spark plugs no longer were necessary with the compact, 2.5-inch bore but the Delco distributor was retained, along with a small eight-volt battery to give starting current and maintain constant voltage. The spark plugs were located on the inlet side of the cylinder head with optimum cooling in mind. The Y-shaped rocker arms were drilled for lightness, which weakened them further, and a multiple-disc clutch in a light, forged flywheel replaced the conical relic. As always, a three-speed transmision was used.

This was the engine that Harry Miller set out to beat, spurred on by Milton, Vail, and others. In 1922 it was said to be developing 114 HP at 4250 RPM. As mentioned in the last chapter, near the expiration of the three-liter formula in 1922 Duesenberg produced and tested a Miller-type cylinder head and then a train of spur gears in place of the shaft-and-bevel system.

These experiments resulted in dramatic improvements, and for the two-liter, 122-cubic-inch formula of 1923 through 1925 Duesenberg made them standard for his racing engines, although the Miller-type head still was detachable from the Duesey block. The Duesey cast-iron block-crankcase unit was replaced by one of aluminum, using steel wet liners. Crankshaft structure remained the same but now the main bearings were plain at the center and 410 ball race at the front and 216 ball race at the rear. Crankpin diameter was increased from 1.817 to 2.25 inches. A tower of spur gears drove the two camshafts, each of which ran in six bearings, numbers two and five being bronze as before but the others now being ball races. Cup-type cam followers were used, and all valves had a diameter of one inch and lift of .3125. Valve timing was 5-38-35-6 and four dual-throat, 1.5-inch Miller carburetors were fitted. The 1925 supercharged engines used a single Winfield carburetor with either 1.75 or two-inch throat. Four valves per cylinder in pent-roof chambers were used in 1923, but in 1924 the hemispherical combustion chamber was adopted, with two valves per cylinder. Bore and stroke of the Duesey 122 were 2.375 by 3.422 inches.

This near-total capitulation to Miller design became essentially complete with the advent of the 91.5-cubic-inch formula of 1926 through 1929, Duesenberg adopting cast-iron blocks, integral cylinder heads, and five main-bearing crankshafts.

The chrome-nickel-molybdenum crankshaft of the Duesey 91 used a 407 ball race at the front and a 214 at the rear. The plain intermediate main bearings were two inches in diameter and 1.125 and 2.25 inches in length, the latter applying to the center main. Crankpin diameter was 1.625, and length was 1.5 inches. The connecting rods were of the same new alloy as the crankshaft and measured 5.5 inches from center to center; those of the 183 had measured eight inches. The pistons were just 2.0625 long, the wrist pins were full-floating and were retained by aluminum plugs in Miller fashion. Both compression ratio and supercharger drive ratio ranged very close to five to one, and a single 1.75-inch Winfield carburetor was fitted. By 1928 and probably earlier the Duesey 91 was peaking at 7,800 RPM.

Simultaneously with the introduction of the DOHC 91 Duesenberg launched an experiment that promised to be as revolutionary as the supercharger (Chapter 22): a supercharged two-cycle engine based on the four-cycle 91 crankcase.

With each reduction in the piston-displacement limit governing American racing there had been a corresponding, compensating increase in crankshaft speeds. With these, of course, valve-train speeds also increased and always introduced new problems.

The basic advantage of the two-cycle engine—one power stroke for every two strokes instead of for every four—had been attractive to racing men for years. But the shortcomings of the principle were well known: while a two-stroke was theoretically capable of burning twice the fuel and developing twice the power of a four-stroke of the same displacement there were severe limitations to getting the fuel into the cylinders and getting the burned gases out efficiently. Duesenberg realized that his recently perfected supercharger could probably overcome these problems and release the great potential inherent in the two-stroke principle. Moreover, with a straight-eight engine and a two-four-two crankshaft, two cylinders could be made to fire simultaneously, while also getting rid of unbalanced couples. And poppet-valve problems would be totally eliminated.

The engine which carried out this thinking was logical and excited a great deal of interest in professional circles. It used rotary intake valves and window exhausts, the rotary valves making up a single long tube similar to those in the Scheel-Frontenac four-cycle engine. The car ran well in practice at Indianapolis, and Ben Jones qualified it in the sixth row at 92.14 MPH. He ran well in the race itself, holding his own for 54 laps, when the engine locked up and he hit the wall. The experts agreed that this was a promising start which merited further development, but the project was not heard of again. In 1926 E. L. Cord had purchased the assets, such as they were, of the remnants of Fred's passenger-car enterprise and had Fred and Augie working full time on the development of the Model J passenger car. It was to be introduced to the public and market at the New York Salon in December of 1928.

The two-cycle engine had the same 2.286 by 2.75 bore and stroke as the four-cycle Duesey 91. The connecting rods were interchangeable, but deflector pistons were used in the two-cycle. The cylinders were cast, like the four-cycle, in two blocks of four cylinders each and were most unusual in that there was a large aluminum plug, almost as wide as the block itself, threaded into the top of each cylinder. This saucer-shaped plug had a boss at the center of its lower surface which threaded into the center of the combustion chamber, while the rest of the plug was in direct contact with the coolant. The spark plug was threaded into the central boss, and the highly

Model A passenger-car engine was a modified version of the competition engine. JERRY GEBBY

Model A Duesenberg Number 1. Note walking-beam straight-eight engine. EDDIE MILLER

In October 1922, the 183 was fitted with DOHC head, but bevel-gear drive was retained. Note "Duesenberg hose clamps." CHARLES LYTLE

conductive aluminum helped to reduce the then-common problems of spark plug malfunction due to overheating. The otherwise conventional Duesenberg supercharger was provided with two outlets, one for each cylinder block. *Automotive Industries* commented:

> It is interesting to think of an eight cylinder engine firing four cylinders per revolution, but it is more interesting to think of a two-cycle construction in which eight cylinders fire every revolution, two of them firing simultaneously. First No. 1 and No. 8 fire, then No. 4 and No. 5, then No. 2 and No. 7, followed by No. 3 and No. 6.
>
> This firing order necessitates special ignition, there being two distributors, each of which has a four-cylinder cam rotating at engine speed, which is twice the speed required for the conventional four-cycle engine distributor.

The engines of the 300-cubic-inch straight-eight race cars were offset one inch to the left of the chassis center line for better balance and tire wear on the speedways, where all cornering was to the left. Starting with the 183's the offset was increased to 1.5 inches. Then, in the two new Duesies for the 1927 season the engine was mounted diagonally in the frame, with its forward end close to the right-hand frame rail. A diagonal drive shaft, leading to a final-drive assembly located close to the left rear wheel, made for a very low seating position for the driver, very low body lines, and a low center of gravity comparable with that of the Miller front drive. Porting was improved, the included angle between the valves was increased in all of the

Duesey engines, and the crankshaft was lightened, including reduction of its intermediate main bearing diameter from two to 1.625 inches. Delco distributor ignition still was used. The changes resulted in an increase of 700 RPM and helped George Souders achieve his Indianapolis victory.

The 91-cubic-inch engine started out with a bore and stroke of 2.286 by 2.75 inches, but in 1928 these dimensions were changed to 2.1875 by 3.0, at which point an output of 160 BHP was announced. Duesenberg was not in a position, as Miller was, to make every possible component on his own premises, and each of the following parts for the 91 came from a different outside supplier: crankshaft forgings, camshafts, pistons, wrist pins, valves, camshaft bearing bushings, valve springs, radiator core, axle shafts, final drive gears, axle springs, shock aborbers, and steering gears, to name a few.

This does not hint in the slightest that Duesenberg race cars were "assembled" machines; to the contrary, they always bristled with originality, with the characteristic Duesey touch. For example, nearly all the race car frames during the Duesey straight-eight period were highly unique. First, forms were made for two frame rails. Then duraluminum (before its advent, so-called "hard aluminum") sheet about one-eighth-inch thick was pounded over the forms, resulting in deep channel-section rails which were as light as they were lacking in stiffness. Then rails of oak were fitted perfectly into these channels. Then the channels were closed by flat aluminum plates which were riveted to half-inch flanges on the channels. Engine, spring, cross-member, and

The four-valve-per-cylinder DOHC head was so successful that the remainder of the classic system was adopted: camshaft drive by spur-gear train. CHARLES LYTLE

other mounts then were bolted solidly to the rails. Perhaps the resulting frames were no lighter than Miller's, which were of cold-rolled mild steel. But they had a resilient quality which contributed to the suspension effect and, with Duesenberg's much softer semi-elliptic springs, gave the cars a much superior ride. The superiority was most marked in dirt-track competition.

The typical Duesenberg frame had four cross members which were built up from eight-gauge sheet steel. The rear cross member was of box section and was also filled with wood. The cross member ahead of it also served to support the torque-tube yoke and to transmit axle torque to the frame. This construction was said to weigh one third less than a similar steel frame with only two cross members.

Some Duesenbergs used conventional spring shackles but, starting in the early 122 period, the spring ends were flat and without eyes. They rode in "shock insulators" which consisted of rubber blocks over and under the ends of the main spring leaves and carried in aluminum housings of teardrop shape which were bolted to the frame. The rear spring leaves were formed with a 180-degree arch which fitted over the rear axle housing in order to eliminate the upsprung weight of conventional spring saddles. The rear axle housings were built-up tubes of nickel steel and were bolted to a light-alloy center section.

For the 1930 season Fred developed some modifications for the Model A engine. One was simply a crankshaft destroked from the standard 5.0 inches to 4.625, in the interest of higher revs and for a displacement of 244 cubic inches as opposed to the

The original four-valve, bevel-drive head for the 183. The rare and magnificent photos also show excellent details of crankcase, hard-aluminum frame before addition of wood filler, and front axle. TED WILSON

Detail of running gear on 1921 GP machine, including hydraulically operated front brake. GB

stock 260. The other change was a major and interesting one: an entirely new cylinder head. Its most remarkable feature was positive closing of the valves by means of an auxiliary roller on each rocker arm, against which the cams also worked for this purpose. This original and successful form of desmodromic valve action bore no resemblance to the old Peugeot and Delage systems. Porting and water-jacketing were modified for racing requirements, and the spark plugs were moved from their near-horizontal position low in the cylinder head to a central position in the top of the head.

The Model A Duesenberg—introduced in 1921—was nothing more nor less than a road version of the 183-cubic-inch race car. *Motor Age* reported:

> After making a thorough and almost piece-by-piece analysis of the constructions and units incorporated in the present Duesenberg stock chassis and that of the racing car, it can be said without contradiction that the race car chassis is about 75 per cent stock.

The article then went on for pages to show in words, drawings, and photographs how literally true this statement was. For example, the race car got its hydraulic brakes from the road car; their brake drums were interchangeable. The road car's rear axle was a scaled-up version of that of the race car; it weighed 230 pounds versus 430 for the Cadillacs of the period. Many of the running-gear forgings were identical in both the race car and the Model A, with the sole difference that more stock was machined from the components which were used for racing. The engines were essentially the same except for dimensional differences and, in the Model A, a conventional, split crankcase and two valves per cylinder.

Although its engine was small in comparison with luxury-car practice of the day, the Model A was both a very fast car and a very economical one; it was efficient. Owners smugly enjoyed astounding their passengers by dropping the shift lever into top gear and pulling away from a dead standstill with utter smoothness.

The Model A came into being when Rolls-Royce was establishing a large new factory at Springfield, Massachusetts. Model A authority David T. Davis said:

I believe that the Duesenberg brothers were trying to build a car that would outdo the Rolls-Royce and appeal to that sort of clientele, without the truck-like features of that marque; a car that would weigh about a ton less, run rings around it, ride better, and, as an extra bonus, give you about twice the gas mileage and be easier to drive and control with its four-wheel hydraulic brakes and cam-and-

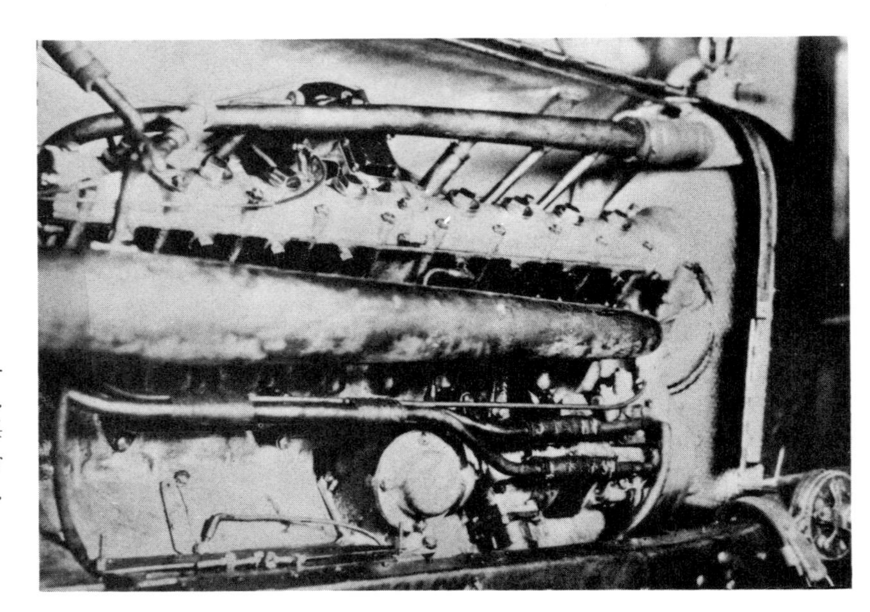

Engine of the 122 was a scaled-down version of the spur-gear 183. This is the Boyer-Corum 1924 "500" engine; note supercharger drive in center of crankcase.
CHARLES LYTLE

The Duesenberg 91 was a scaled-down 122, but for the first year of the new formula a try was made with a supercharged two-cycle based on the four-cycle 91 crankcase.
CHARLES LYTLE

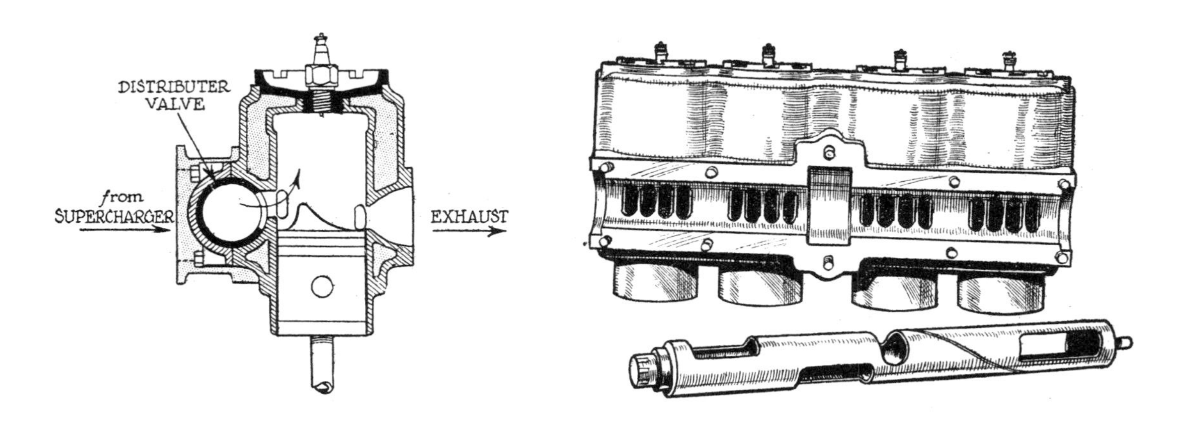

These drawings clarify the structure of the two-cycle's rotary intake and window exhaust valves as well as the peculiar plug in the top of the cylinder head. *Autocar,* LONDON

lever steering; a car with a high-speed, high-performance, small-bore engine instead of the outmoded slow-speed fire-truck job used by Rolls.

Expert William A. Johnson has determined that about 667 Model A's were built, the last 17 being called Model X but differing from the A only in details of appearance. These figures are interesting in that they show that far more A's were built than the approximately 470 much more famous Models J and SJ. How did the two compare? Van Ranst was close to both projects:

> E. L. Cord told Duesey what he wanted. Fred of course always thought in terms of high-speed, high-output engines. He had never concerned himself with the necessity for high torque for getaway and acceleration. So Fred designed the prototype J in the same spirit that he designed the A. He gave it a little over 300 cubic inches' displacement and the performance and size of the car didn't suit E. L. at all. Fred enlarged everything but E. L. still wasn't satisfied and insisted that it be enlarged again. It wound up at 420 cubic inches and all this rehashing left the J quite unlike what Fred had intended it to be. Low-end torque and power-to-weight remained poor in spite of all the inches and the J still had no real acceleration, in spite of its 116 MPH advertised top speed.

Supercharging helped the J's performance to the tune of a claimed 104 MPH in second gear and zero to 100 MPH in 17 seconds—substantially beyond the safe capacity of the best passenger-car tires of the time. The car was every inch a thoroughbred in the Duesenberg racing tradition, and it was much more. It was every inch a connoisseur's machine, and it ranks with the handful of most magnificient automobiles of all time. With all due credit to E. L. Cord, it is a fitting monument to Fred and Augie, who lived for and did so much to create the American thoroughbred car.

chapter 16

FRONTENAC: THE YEARS OF GLORY

FRONTENAC'S TRULY BRILLIANT racing record was noted by William Small, the top man of the Monroe Motor Company of Flint and Indianapolis which, incidentally, was linked closely with the Durant empire. Small wanted a team of four cars to carry the Monroe name during the coming racing season, and at Indianapolis in 1919 he persuaded Louis Chevrolet to create them and to move from his Plainfield, New Jersey, job with American Motors to Indianapolis to do so. He also agreed that Louis should be free to build duplicate cars, at Louis' own expense, to run under the Frontenac name. Louis, thoroughly impressed by van Ranst's ability, journeyed to New Jersey and offered Van a consulting role in the new venture. Van's aircraft engine project had just gone the way of all wartime industry and he welcomed the new opportunity. Small provided a well-equipped shop in which to build the cars—four green Monroes and three Burgundy-red Fronties. Van modernized Etienne Planche's original cast-iron DOHC four, and all seven of the new cars qualified for the 1920 "500"—a nice achievement in itself. They were light and fleet and undoubtedly the finest all-American machines that ever had raced.

Just as this project was being worked out on the drawing board Henry Ford's metallurgical wizard, Childe Harold Wills, had rocked the engineering world with the remarkable properties of chrome-vanadium steel. This breakthrough made possible great reductions in the weight of automotive components and made the Model T Ford the lightest car on the market as well as one of the strongest. Louis became an immediate convert to chrome-vanadium steel.

Flower-petal radiator shells distinguish the Frontenacs and Monroes in the 1920 Indianapolis starting line-up. GB COLLECTION

He bought his I-beam front axles, rear axle housing, and various running-gear parts from the Colombia Axle Company. This firm had been supplying him with steering arms of chrome-nickel steel, which was extremely adequate, but he specified the new alloy for his new cars. Time went by, the cars were assembled, but the steering arms still had not arrived. Louis called the Cleveland factory in protest but was told that such special materials naturally required special handling and time.

However, an expediter was sent into the forging plant to speed the job along. He found the parts, all neatly tagged, and sent them to the machine shop for finishing, blissfully ignorant of the fact that they had not been heat-treated.

So the steering arms were received, installed, and they broke in the order of assembly, after about 550 miles of running. None broke in practice because there wasn't that much practice.

Art Klein got the first car and got in about 400 miles of practice. He lasted in the race for forty laps, when his steering arm broke and he hit the wall. Eighteen laps later Roscoe Sarles dropped out from the same cause. The next victim of the epidemic was Louis, on Lap 94. Bennett Hill came into the pits, and as Sarles took over to drive his relief, Barney Oldfield shouted, "Hey, Roscoe! Do it again! I didn't get to see it."

And sure enough, on Lap 115 that steering arm let go and Sarles ground into the wall.

Joe Boyer, second fastest qualifier in the field at 96.90 MPH, led the race for most of the first 250 miles, had his steering arm break on Lap 192, but still collected twelfth place and 9,500 dollars in lap prizes.

Gaston Chevrolet had been so busy with his duties as superintendent of the little Fronty plant that he took the last car to be assembled and had no time for practice. He held a steady pace which kept him in second spot almost until the end. Then the leader, de Palma in a Ballot, ran out of fuel far from the pits and Gaston rolled on to victory. He was not the best driver on his team, and if the other cars had stayed in one piece the whole Fronty flotilla might have swept the field.

During this race Louis' and Albert Champion's fates crossed again. Louis had had many sad experiences with ceramic insulator spark plugs breaking up, getting under the valves, and causing grave damage; for this reason he had come to depend upon Mosier mica plugs. Bob Stranahan, who had purchased the Champion Spark Plug Co., knew of Louis' hostility toward the name and so sought out van Ranst with the news that his firm had developed a ceramic insulator that was vastly better than the Frenchtown porcelain that was standard at the time. He begged Van to test them at least and added that there was 5,000 dollars in extra prize money for anyone who could win the big race with these plugs. Also, Delco had developed its new light and compact battery and distributor ignition to a high level of reliability and made an identical cash offer. The little enterprise needed every cent it could get; Van was persuaded and he in turn persuaded Gaston to try the new equipment while sparing Louis the bothersome facts.

With six of the seven new cars out of the race and Gaston still going, Van was knitting buttonholes and fervently praying that the unauthorized equipment would not fail. When Louis learned the truth and realized that he had just put Champion in the racing business he was like a raging lion. "Idiots!" he shouted. "How could you dare? That junk might have failed!"

"You mean like your damned steering arms?" Gaston snapped back.

Louis, beyond words, gave one of the front tires of Gaston's car a savage kick. Its steering arm fell to the floor with a clang that ended the argument.

The victory and the unexpected 10,000 dollars in prize money healed the wounds soon enough. In the immediate postwar period American nationalism was at a peak and this, the first win by an American car since 1912, thrilled the entire country. Louis and Gaston were great heroes overnight. But all was not joy in the Chevrolet camp, and what joy there was did not last long.

During the practice for this race René Thomas, in one of the new Ballots, blew a tire and spun in front of Arthur Chevrolet's Monroe when both were traveling at about 95 MPH. The two cars collided, the Monroe overturned and slid for hundreds of feet with Art trapped in the cockpit by the novel steering wheel which had to be mounted on the steering column after the driver had squeezed into his seat. He was severely mangled and barely survived.

Then, on November 25 at the Beverly Hills board speedway, Gaston's Monroe crashed, and he was instantly killed. But in the midst of all this disaster and although

The 1920 winners: Gaston Chevrolet and Monroe. GB COLLECTION

his cars for the new formula were wonderfully fast, Louis rushed into the creation of an entirely new engine—a 183-cubic-inch straight eight. He had the soundest of motives: the Ballot eights were faster than his fours.

Van Ranst engineered the new power plant in the late summer and fall of 1920, and two were completed in time for Indianapolis qualifications in 1921. Seasoned Fronty driver Ralph Mulford was assigned one of the cars, but Louis needed a top driver for the other machine. Barney Oldfield was aware of Tommy Milton's agonies with Duesenberg and Miller, advised that he was the hottest thing on four wheels, and Louis signed him on.

In practice the new eights stayed together and ran like watches up to the old, tried-and-true rev limit of about 3,000 RPM. But the instant that figure was exceeded, out would go a connecting-rod bearing. Frenzied changes were made but always with the same result until the crankcases were a mass of patches and were spewing oil like sieves. With time absolutely running out, Milton went out to qualify on the Saturday before Monday's race. He knew that 3,000 was safe and that it should be fast enough to get into the race, which he did. Then, to see if the latest change had helped the problem, he floor-boarded the throttle and took the engine out to its 3,300 RPM peak. *Thunk!* and out went another rod. The engine was punched full of new holes when the crew rushed back to the plant at five in the afternoon. It had to be rebuilt and running by Monday dawn at the latest.

In those days the market was flooded with "miracle" bearing metals for which fantastic claims were made—they would run without oil, for example. But which of

Louis Chevrolet during an early test of the Scheel-Frontenac. Engine has window-type exhaust valves on both sides, giving impression of a V8 exhaust system. Old Speedway pagoda is in background. IMS

these marvels to use? Mulford, a very serious, sincere, and religious person, said, "Louis, I have an idea. Why don't you put a shaft in the lathe and run a life test on all these special metals?"

Louis drew a deep breath, turned slowly, and, in his still-strong French accent, said, "Ralph, you really want to help, don't you?"

"But of course I do, Louie. You know that."

"Then *please* just go home," his boss pleaded.

The hours rushed by and race day dawned. The engine had been slammed together in the wildest haste, and it oozed oil from a hundred seams, too many of which were right over the clutch. This is how the Great Milton remembered his first Indianapolis victory:

> I took off as well as I could but the clutch was slipping horribly and I couldn't begin to approach any really useful revs. For countless laps as I came off every corner I'd punch the throttle and let the clutch slip for a couple of

seconds, then back off. My plan of course was to burn the oil off the clutch until it began to take hold, which it actually, finally did.

At about that time a spark plug went sour. There were two of the things for every cylinder and if you hung by your heels and had a socket wrench with eighteen U-joints in it, you could get them out. However, with that one plug not firing I found that I had an automatic governor that limited the engine revs to the very edge of the safe operating range. Mulford had the same trouble and stopped and changed a plug or two but I reasoned that there was no sense in stopping if you could run on seven and a half cylinders and turn 3,000 RPM anyway.

The oil loss got worse as the race wore on and I had to stop for oil and tires at the half-way point and then again at about 400 miles. Roscoe Sarles was driving for Duesey and had a car that was a lot faster than mine, and when I made that last pit stop he sailed into first place. I told myself that I had to do something to beat him mentally. If it ever got to be a race I would be a dead pigeon.

I remembered how Mulford had beaten me at Sheepshead Bay once and decided to use the same tactic. The Indy track was all brick and it was streaming with oil. They made it a practice to throw sand on the oil and it always was a moot point which was more slippery, the oil or the sand. Anyway, I picked the sandiest, slipperiest turn on the track to pass poor old Roscoe in. I drove into it much faster than I dared, fishtailing madly. As I streaked and wobbled past him at fearsome speed I took one hand off the wheel and gave him a jaunty wave. Then I looked back at him, grinning as though I was having the damndest good time ever, and pounded on the tail of the car like a horse jockey hungry for more speed. It worked, and Roscoe decided that he didn't want to be anywhere near such a maniac. As we shot past the pits Fred Duesenberg was doing everything but throwing buckets at Roscoe to make him go. He just gave Duesey the long nose, and I won the race.

It was a great day for Louis Chevrolet. In addition to Milton's first place, Jules Ellingboe in a Fronty four finished third, and Mulford, in the other eight, was ninth. Only nine out of twenty-three starters went the distance.

It was a significant day in automotive history since it marked the first victory by a straight eight in one of the world's major contests. Duesenberg eights had scored many wins in secondary events, and this time at Indianapolis de Palma's Ballot eight again was the fastest qualifier at 100.75 MPH. In the race he set a flying pace and led for all but two of the first 112 laps—and then threw a connecting rod. Ora Haibe, in a Sunbeam straight eight, came in fifth, best of the six European entries.

Milton had decided the night before the event that his car was too tight and replaced thirty-two of its bearings, finishing his labor at two in the morning of the big day. He drove a steady, heady 500 miles at an average of 89.62—just .22 MPH slower than the record which de Palma had set with his big, 300-cubic-inch Mercedes

The Frontenac straight eight. Out of eleven four- and eight-cylinder entries in the 1922 "500," the marque's best placing was ninth. IMS

six years before. For the entire racing world this was crowning confirmation of the superiority of eight cylinders over four.

Here the Fronty story flashes back to action-packed 1920. Dirt-track racing had revived as a hugely popular spectator sport. But it was expensive because race cars were expensive, they wore out fast on the dirt, they collided often, and every time they did it cost a fat sum to repair them. But the market was becoming flooded with all manner of racy conversion parts for the Model T Ford: speedster bodies, special running-gear components, and even "Peugeot-type" Laurel-Roof OHV cylinder heads. The latter were not too successful, but the idea was a good one. So one day Van braced Louis:

> I think you can make more money on the dirt tracks with cheap cars than you ever can with good cars. No one cares whether they do 60 or 100 MPH as long as they're all bunched together, moving at a good clip, do a little bumping and kick up a lot of noise and dust. We could make up a bunch of cars using Hassler's good running-gear conversions and Morton & Brett's speedster bodies and they'd put on a terrific show. And we can design an OHV head for the Model T that will run circles around the Roof.

The little Chevrolet Brothers Manufacturing Company at 410 West Tenth Street in Indianapolis needed extra work at the time. Louis liked the idea and put Van to work on the design of the original eight-valve, pushrod Fronty head. As the work neared completion Van looked around for the nearest Model T to try it on. It turned out to belong to shop-welder and parts-chaser Skinny Clemons, who later won note as a dirt-track driver, then race-car owner, then Wilbur Shaw's racing mechanic, and then a builder of racing engines in his own right. Van offered Clemons the first head as a gift in exchange for the use of his T, and Clemons jumped at the offer.

The day came when the first prototype head was torqued down on the block of Clemon's modest little two-door sedan. Van and Louis took off with Louis at the wheel. It had been years since either had driven a Model T—even a normal one—and this one was by far the hottest T that had ever hit the road. They headed for the countryside, looking for a good, paved test course. Indiana's rural roads in those days were surfaced with limestone gravel, and it was hard to tell from a distance where concrete stopped and gravel began. The one clue for the sharp-eyed was that the cars used to grind the gravel into a fine powder which road crews periodically scraped and piled in the middle of the road for the cars to spread out again.

Louis was hurtling down the concrete pavement flat-out at what seemed like at at least 100 MPH in the high, tottering T, when the pavement suddenly ended without visible warning. The left front wheel dug into the loose rubble in the center of the road and, in a flash, the black, boxy sedan was skating upside down at furious speed. Louis was on top of Van, both were afraid that the car would burst into flames at any moment and were doing their best to kick the doors open. One of them finally succeeded, and both tumbled out on top of each other, miraculously unscratched.

They stood there, stunned, staring at the still-spinning wooden spoke wheels, when a group of round-eyed kids who had witnessed the whole drama from their Essex touring car stopped and helped to put the T-bone right side up. Its top had been ground all the way through its wooden slats: the price that Clemons had to pay for the first-ever Fronty head.

"It's fast enough for a race car," said Louis, and they tooled sedately back to Indianapolis and built five dirt-track machines. The Fronty head was an immediate success and unsolicited orders for it began pouring in from the racing fraternity and then from the public at large. Soon Louis was cranking them out at the rate of sixty per day, and he sold well over 10,000 of these heads before the interest in souping up Model T's finally trickled out. He was one of the true founding fathers of the tremendous speed equipment industry that lay in the future.

Fronty-Ford race cars confirmed van Ranst's vision by quickly revolutionizing competition on the dirt. Also, they did remarkably well in the big time. In the 1922 "500" two were entered, driven by John Curtner and Glenn Howard. They still were

Henry Ford at the wheel of one of the remarkable Barber-Warnock Fronties, one of which was driven by Sterling Moss's father, then a Fronty agent in England. Behind Ford are Louis Chevrolet and Barney Oldfield. FORD MOTOR CO.

in the infancy of their development and were the slowest cars in the race, in spite of Curtner's perfect reliability, until he was flagged at 160 laps, and Howard's, for the entire distance.

In that 1922 race there were nine other entries for which Louis Chevrolet was responsible, in addition to the Fronty Fords. Two were four-cylinder Frontenacs which he and Van designed specifically for this event and which used the top-end layout of the Fronty straight eight. But eights dominated the race, and Tom Alley's Fronty Monroe was the only four-banger to finish in the money, coming in ninth.

For 1923 Louis converted one of the latest "straight-eight-type" fours to a rotary-valve system which had been designed by wealthy St. Louis engineer Herbert Scheel. Four Scheel-Frontenacs were entered in the "500" but only the one actually was built. During a shakedown run with Scheel himself driving, the car caught fire on the Indianapolis backstretch and was a total loss. The project was abandoned.

The biggest surprise of the 1923 "500" was a Fronty Ford which was named the Barber-Warnock Special, for the Indianapolis Ford agency which sponsored it. L. L. Corum drove the little car with spectacular consistency, went the entire distance without relief, without a tire change, and his was the last car to stop for fuel. It ran away from the Mercedes and Bugatti team entries, as well as most of the best American machines, and shot across the finish fifth overall. Corum qualified the noisy four-banger at a healthy 86.92 MPH and averaged 82.58 for the 500 miles, just 8.37 MPH slower for the distance than Milton's winning Miller HCS Special. According to one publication of the day, when Corum crossed the finish line he drew the loudest and longest roar of applause ever heard at the Speedway. After all, out of the 150,000

Left: Louis Chevrolet and the historic Buick *Bug*, Indianapolis, 1910. MRS. SUZANNE CHEVROLET Right: Arthur Chevrolet at the wheel of a racing Marquette Buick, Indianapolis, 1911. KARL LUDVIGSEN

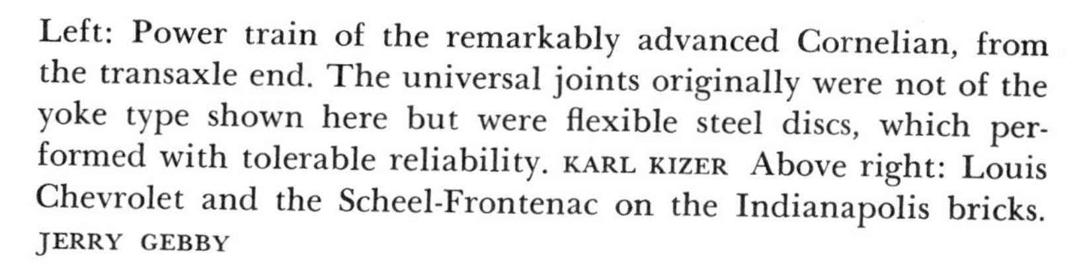

Left: Power train of the remarkably advanced Cornelian, from the transaxle end. The universal joints originally were not of the yoke type shown here but were flexible steel discs, which performed with tolerable reliability. KARL KIZER Above right: Louis Chevrolet and the Scheel-Frontenac on the Indianapolis bricks. JERRY GEBBY

spectators there were more than a few passionately partisan Model T owners and boosters, and they all roared themselves hoarse every time the steady little underdog ticked off another lap. Arthur Chevrolet had prepared the car, and its remarkable performance made him one of the heroes of the year.

This success, hailed as a great moral victory, fired Barber-Warnock to commission the construction of three similar cars for the 1924 "500." It may have been no coincidence that old Henry Ford himself agreed to serve as referee for that year's race and, while there were no indications that his company was directly backing the Fronties, he took an understandably lively interest in them, unbending to the extent of permitting himself to be photographed in one.

But huge strides had been made by the builders of the thoroughbreds during the preceding twelve months. The orthodox straight eights were developing in the neighborhood of 120 BHP, versus 80 for the hottest Fronty, on top of which Duesenberg had just cracked the combination to supercharging. The result was a rout for the Fronty Fords. Even though they ran with their previous consistency, no drivers had to be relieved, and their speed was enough to have put them all in the top ten in the previous year's race. A pleasant little footnote to history is that one of the Barber-Warnock drivers was a British importer of Fronty speed equipment named A. E. Moss. Five years later his wife would bear him a son, and he would name him Stirling.

For the 1926 season Louis built a new Fronty Ford, this one a front-wheel drive called the Hamlin Special. To bring it within the 91-cubic-inch limit its bore and stroke were 2.875 by 3.5 inches, achieved by sleeving the stock Ford block and

The Monroe in which Gaston Chevrolet won at Indianapolis and lost his life at Beverly Hills. ERNIE OLSON

machining a short-throw crankshaft. The engine was installed in the frame with its flywheel end forward while, at the rear, a Roots supercharger was gear driven off the nose of the crankshaft. The compression ratio was 6.75 to one, all the critical parts were drilled for pressure lubrication as with all Fronty Fords and peak revs were 6,000, to the delirious inspiration of all Model T owners. And, as with the previous Fronty Fords, about seventy per cent of the car's components were right out of the local Ford dealer's parts bins.

Jack McCarver qualified the Hamlin Special in the ninth row—ahead of de Paolo's Duesenberg!—at Indianapolis in 1926 but lost a connecting rod bearing on Lap 22. But as late as 1932 this car still was a money-maker in the minor league under the name of the Ray Day Piston Special.

Famous driver Chet Miller had his first go at Indianapolis in 1930 in a Fronty Ford prepared by Arthur Chevrolet. *Motor* reported:

> A racing car that can be bought retail for $2,500 was the Fronty Ford consisting of Ford Model T parts except for Fronty cylinder head, pistons, rods, crankshaft and rear axle shafts. Chet Miller set out to average 100 MPH and but for minor

Engineer Scheel, right, studies the power plant of the Scheel-Frontenac. JERRY GEBBY

The Frontenac-Stutz prototype was built in 1922. Note the front-wheel brakes. MRS. SUZANNE CHEVROLET

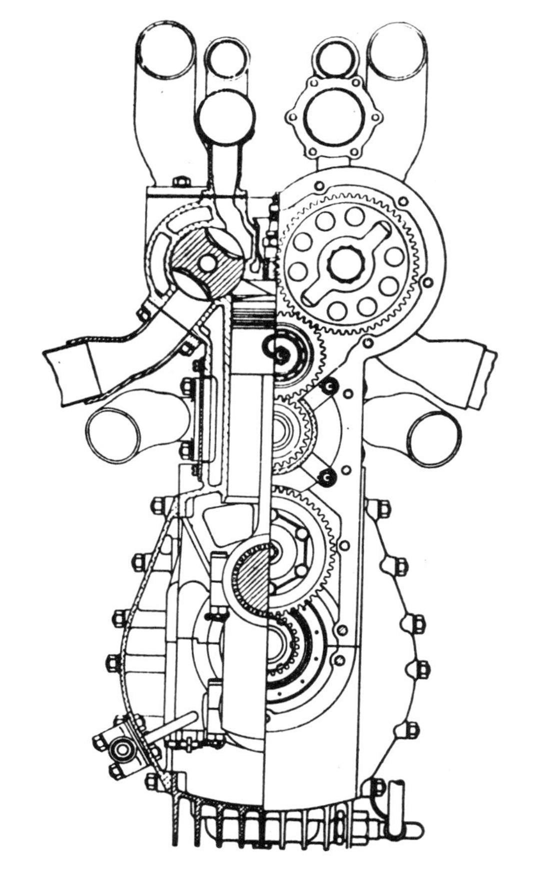

Transverse cross-section of the Scheel-Frontenac engine. Peugeot-type gear train drove rotary intake valves. Crankcase was barrel type and crankshaft rode on roller bearings. GB COLLECTION

mechanical troubles might have attained this goal. As it was, if a 41-minute stop to replace a broken front spring is deducted, his average becomes 96 MPH.

The replacement of this spring is one of the funniest happenings heard at a race in many years. Chet Miller swears it is true. When he stopped on the 92nd lap to adjust the carburetor, the technical committee discovered the front spring broken close to the spring eye, and refused to let him proceed until it was replaced. Lacking a spare spring, Miller and his mechanics ran back into the infield, found a Model T Ford with no owner nearby, removed the spring, put it in the race car and continued the race until flagged. Then the spring was removed and returned to the Ford before the owner discovered what had happened.

The following year Gene Haustein lasted for twenty-three laps in a similar car and then threw a wheel, ending Frontenac's career in the big time. In the minor leagues the name remained a magic and formidable one for years to come.

chapter 17

FRONTENAC: TECHNICAL NOTES

ALL OF LOUIS CHEVROLET'S RACING SUCCESS up to the time he was approached by Monroe in 1919 had been obtained with his SOHC aluminum engine, which had been laid out originally in 1915. The engine was quite powerful, but the aluminum alloys of the period were porous, weak, and unreliable. Also, in the interim the superiority of dual overhead camshafts for ultimate-output engines had been established beyond all doubt. So, when Monroe contracted for a team of new world-beaters, Louis reconsidered the design of the iron DOHC 300-cubic-inch engine which Etienne Planche had laid out for him in 1914. Van Ranst concurred that it was essentially what they needed and proceeded to scale the Peugeot-like design down to 183 cubic inches and to add all known up-to-the-instant refinements. The result was remarkably tiny and light; nearly a toy.

The new engine had a bore and stroke of 3.125 by 5.9375 inches and a displacement of 182.5. Its four valves per cylinder were inclined at an included angle of about 38 degrees, were actuated by finger-type cam followers, and were closed by concentric pairs of coil springs. These were exposed to the air while each of the two camshaft assemblies was carried in its own oil-tight aluminum housing in the Peugeot manner. Camshaft drive was by a train of spur gears.

The cylinder block, head, and the upper portion of the crankcase all were a single, integral iron casting. The crankcase itself was a barrel-section aluminum structure. The two-piece, three-ball-bearing crankshaft was retained, and huge through-

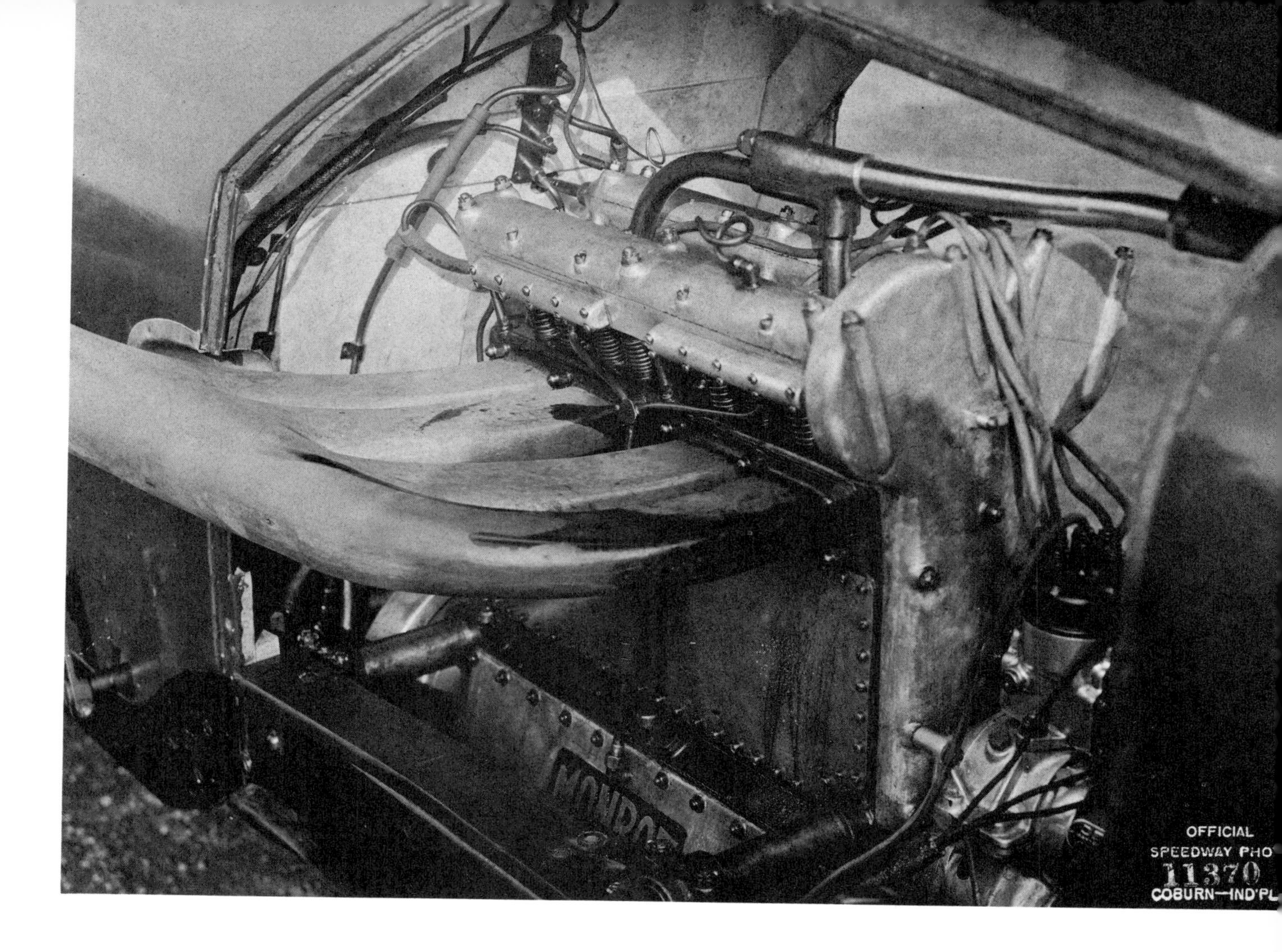

The winning engine of the 1920 "500." VAN RANST

bolts ran from the main-bearing caps to the lower flange of the cylinder block. The successful dry-sump lubrication system of the 300-cubic-inch Fronties also was carried over. The previous cone clutch was replaced by a multiple-disc type, final drive was three to one, tires (Oldfield) measured 32 by 4.5, and wheelbase was 98 inches. This was the first American winner of the Indianapolis "500" since 1912.

The Indianapolis-winning straight eight of 1921 was inspired very frankly by Ballot. It used Ballot's cup-type cam followers and completely enclosed valve gear. It also used a four-four crankshaft with five main bearings, thereby, too, anticipating Duesenberg and of course Miller. The one-piece crankshaft ran in plain bearings, and the engine boasted full pressure lubrication on the dry-sump principle. After Gaston Chevrolet's success with the Delco distributor in 1920 the straight eight was designed to employ this ignition system. The structure and materials of crankcase and cylinder block were basically the same as those of the 1920 four cylinder and made for an excellent combination of strength with light weight. Bore and stroke were 2.625 by 4.219 inches, and displacement again was 182.5. A "double-drop" channel

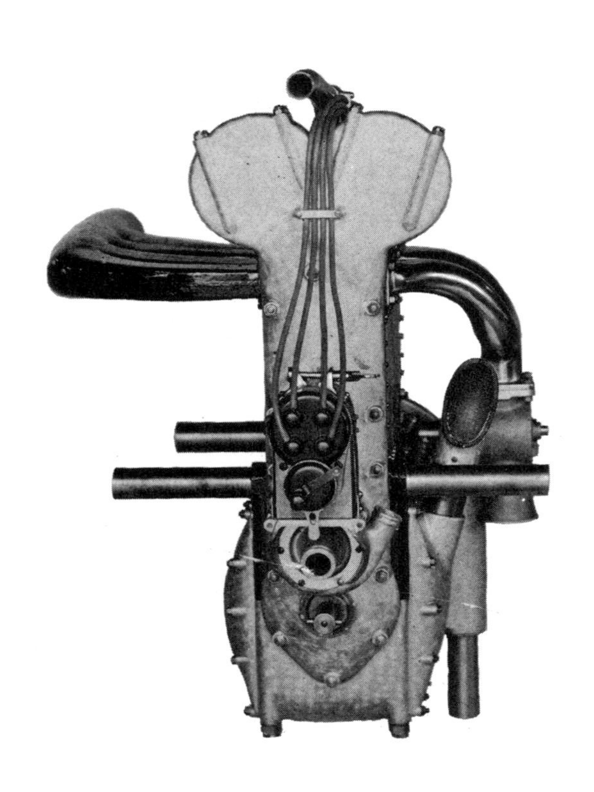

Left: frontal view of the Frontenac-Monroe 183. VAN RANST   Right: Cylinder block of same engine. Note supports for large main-bearing ball-races. VAN RANST

section frame was used. Its height was conventional from the frame horns aft to the firewall; then it dropped about four inches and remained on that level until it kicked up over the rear axle. The purpose of course was to seat the driver and mechanic low in the interest of a low center of gravity. The final drive ratio was 3.25 to one, wheelbase was 102 inches, and the car's dry weight was a very light 1,850 pounds.

Louis still was not convinced that, given the same top-end refinements, four cylinders could not do just as good a job as eight and for the 1922 Indianapolis race he and van Ranst produced a pair of four-cylinder cars which used the eight's valve mechanism and double-drop frame. But even eleven entries could not offset Frontenac's bad luck that year, and only Tom Alley squeaked into the money in ninth place. His mount was one of the old Monroes which, thanks to its novel disc wheels, could run ten PSI more inflation pressure than the competition's Rudge-Whitworths were recommended to bear. Following this 1922 race Louis became totally preoccupied with other projects. One was his promising Stutz-Frontenac sports car. The other was the suddenly booming demand for the Fronty heads. Van Ranst left him in December to join the engineering staff of the Paige-Detroit Motor Car Company.

The legendary Fronty head came in a variety of forms *(see Model T Ford in Speed and Sport,* Post Publications, Arcadia, California, for massive documentation). The original was called the Model R—probably for van Ranst—and was the type with which Curtner and Howard made history at Indianapolis in 1922. Like all the others it was made of close-grain semi-steel. It had one inlet and three exhaust ports, each with a diameter of 1.625 inches. There were two valves per cylinder, made of high-tungsten steel and 1.875 inches in diameter. They were operated by drop-forged, case-hardened rocker arms and adjustable pushrods. With standard pistons compression pressure was 85 PSI.

This outright racing head was followed by the Model S, detuned for what would be called sports use today. It differed from the Model R chiefly in its 1.8125-inch valves and 75 PSI compression pressure. Then, for the mass market, there was the Model T. Its 60 PSI compression pressure was the highest that the pump fuel of the period would tolerate. A stock Model T Ford engine pulled 17 BHP on a Purdue University dynamometer. The same engine, equipped with a Model T Fronty head, yielded 33 BHP. It was this nearly doubled output, for just $98.75, that kept production manager Arthur Chevrolet working halfway around the clock every day for several years.

When van Ranst left Louis he left behind him the design for the Model SR head, which was created to beat the Model R. It had two 1.75-inch intake ports and thus could mount two big carburetors. It also had two spark plugs per cylinder—one on each side—and 100 PSI compression pressure. This was the head that enabled Corum to qualify at 86.92 MPH at Indianapolis in 1923 and to finish an amazing fifth

The Frontenac straight eight with Delco distributor. VAN RANST

The straight eight in Milton's car, winner of the 1921 "500." VAN RANST

Frontenac pushrod OHV conversion made fast cars of Model T Fords. GB

The Chevrolair aero engine was an air-cooled version of the classic racing four. MRS. SUZANNE CHEVROLET

over-all. The following year Frank Lockhart found the price of an SR head and began shattering dirt-track records on the West Coast.

Then there was the Scheel-Frontenac interlude of 1923. Herbert Scheel had developed an original approach to two-cycle engine operation. Among its features were long, tubular, rotary inlet valves which gave a multicylinder engine the appearance of a DOHC four-stroke. It also had window-type exhaust ports low on *both* sides of each cylinder wall, plus a carburetor for each cylinder. The carburetors did not feed their charge into the crankcase in typical two-stroke style but instead into the space between the underside of the piston and a diaphragm which sealed the cylinder from the crankcase. Louis saw enough merit in the idea to make a Scheel-Fronty out of one of his 1922 "enclosed valve" fours. It was probably a crankcase explosion that put an end to the experiment.

On the heels of this, Louis and Arthur produced a 16-valve DOHC head for the Ford T. This, called the Model DO, had an inlet and exhaust port for each cylinder. One spark plug was located centrally in the top of each combustion chamber. The valves were 1.5625 inches in diameter, their included angle was 60 degrees and the cam followers seem to have been of the mushroom Hispano-Suiza type—discs threaded onto the tops of the valve stems. Camshaft drive was by means of link-belt chain and the compression pressure was 120 PSI.

The Barber-Warnock entries for 1924 had bores and strokes of 3.115 by 4.0 inches and displacements just under 122. Two of these cars ran milder Fronty heads, 6.5 to one compression ratios and developed 65 BHP at 3,700 RPM. The third ran a Fronty DO head, 7.0 to one compression, pulled 80 BHP and averaged 88 MPH for the last 300 miles of the race.

Finally, wringing the last out of the Model T Ford as the basis for a racing engine, in 1928 Arthur Chevrolet developed SOHC conversions for the Fronty R and SR heads. These used flat, cup-type cam followers which were free to rotate in their guides. It was with one of these heads that Chet Miller hoped to average 100 MPH at Indianapolis for 500 miles—the hope that was shattered by the need to steal a front cross-spring from a spectator's Model T.

chapter 18

MILLER: THE GOSPEL ACCORDING TO MILTON

"IT WAS IN 1919 that Miller built his first four-cylinder 183. Cliff Durant sponsored the project and it cost him $27,000 but he was so disappointed with the machine that he literally gave it to Tommy Milton. They built a new straight-eight engine for that car after Milton took it over. That was the beginning of Miller's success with racing equipment."—Ed Winfield

Milton told the story in later years with his usual authority and color:

> Of course I broke with Duesenberg over the Land Speed Record incident at Daytona and I went looking for other rides. In the summer of 1920 I made a deal with Cliff Durant to drive his Miller-built car in the Elgin, Illinois, Road Race. This Durant Special chassis was identical to that of Barney Oldfield's Golden Submarine, including the desmodromic valves. It wouldn't go fast enough to blow your hat off. Cliff bought a 183 straight eight from Duesenberg for the machine and, with his typical generosity and enthusiasm, loaned me the car for as long as I wanted to campaign it. I struggled with it in a couple of races and at Beverly Hills I button-holed Fred and said, "Look, this is a lousy engine that you sold Durant and either you're going to give us a new one or I'm going to build one that will go." I demanded that he give it to us free. He agreed but Mickey, his wife, talked him out of it on the way back East.

Tommy Milton and Harry C. Stutz, 1923. IMS

I was headquartered at Miller's and I discussed new engine possibilities with him. I thought that I knew all there was to know about racing engines and my idea wasn't too illogical at that. It incorporated all the best features of all the racing engines about which we knew anything. Among other things, I wanted to copy the Peugeot top end and the Duesey crank with the Peugeot/Ballot rings and passages around the crank cheeks so you didn't have to go against centrifugal force to get the oil to the crankpins.

Miller agreed to tackle the job for $4,000 but warned that it might run a thousand more. Oldfield had told me many times that if I ever needed money I should call on him. I did and he said, "Man, you've picked a bad time. Firestone have just passed a dividend and all my eggs are in their basket." Both he and Cliff told me that I was crazy and that they knew from experience that Miller couldn't build a rat trap. "Don't you think I know that?" I said. "But *I'm* going to dictate this design." Barney shrugged and actually went and borrowed money from his bank to loan me. The project was no sooner under way than Ira Vail said he'd buy one of the new engines, which helped to keep the cost down.

Leo Goossen went to work on the design and I worked with him on the drawing board. The engine got built and put in the car and we took it out to the Beverly Hills Speedway for the first big test. It had no speed at all and within ten miles all the exhaust valves were warped. We fought with the problem and I took the car to Fresno in the spring of '21 but it was so miserable that I left the car out West and went back and drove for Louis Chevrolet. And that eight-cylinder Fronty wasn't much better, even though de Palma let us win Indianapolis with it, almost by default. Ira Vail's experience paralleled mine and he also left his new Miller on the Coast.

I drove a Duesey at Uniontown, beating Murphy to win, and then went back to California to get back to work on the Miller. The valve warpage problem had not improved and we were at our wits' end when we finally modified the cooling system by putting baffles in the block to re-route the circulation and by using discharge plugs with a much smaller bore, so that a pressure of about two PSI was maintained on the water in the block. This kept the water from just boiling away over the combustion chambers and cured the warping of the valves. *That* touch I contributed personally. But we still were getting no power.

I was pretty sore with Duesenberg. I had a little money while I was racing and I used to pay the freight. Fred didn't have enough money to keep the race cars going and I would let him settle up with me at the end of each year. I though I'd been very fair with Duesey and I didn't think that he'd been very fair with me.

Now if you ever worked for a boss and were dissatisfied with him you know that in all probability you weren't the only one. I phoned one of his men and said, "I don't know where your sympathies lie, but I think Fred has given me a pretty bad time and I don't think he's been too good to you. I'd like to get the dope on this new Hall-Scott camshaft that's going so good." He said, "Buddy, I'm with you and I'll mail you a print."

The thing was so revolutionary compared with what we had been using that I was unwilling to believe the direction of rotation that was indicated on the drawing; I thought the draftsman must have made a mistake. Well, Duesey had his cars in Los Angeles and Roscoe Sarles, a good friend of mine, had his car at a Roamer agency there. I went and said to Roscoe, "You know damned well that I can't steal anything just by looking at your engine, don't you? Will you just take the cam cover off and turn the crank once so I can see how this thing rotates?" Roscoe swung the quick-release lever on the cam cover and I saw that the print was correct. I rushed back to Miller's.

The first man actually to buy a Miller straight-eight 183 was Ira Vail and when he left that car on the Coast he left it for Frank Elliott to drive. Frank's brother was a mathematics professor at Caltech and they, too, figured that the original Miller cams were wrong and came up with their own new design. It was no improvement. So I turned the unidentified print over to Miller and paid handsomely to have the new cams made. They were for roller tappets of course and Goossen had to translate the contours for use with radiussed cups. He

While Milton was campaigning the HCS Miller, the author dreamed behind the wheel of his mother's HCS. GB COLLECTION

apparently did a perfect job because my car immediately began running like a spotted-assed ape. Then, on July 4 at Tacoma and in front of 30,000 people, I won the big 250-mile event and its $25,000 purse. The news was out. Both Murphy and Harry Hartz rushed orders to Miller for new engines for their Duesenberg French Grand Prix chassis.

Out of gratitude to Cliff, who had loaned me his car at no cost, I had registered it with the Contest Board as a Durant Special. His father was out of General Motors again and I hadn't paid much attention to the fact that he was organizing the Durant Motor Co. Well, in the spring of '22 there were some sprint races at Beverly Hills and I won them in a breeze and set several new world's records. The new Durant passenger car had just been, or was just about to be introduced and a dealer in Illinois ran an ad in a Chicago daily that read, "Milton sets all new world's record in a Durant Four." That was when I got the worst deal in AAA history.

The Contest Board had very strict rules about misrepresentation in advertising and they promptly advised me, "Milton, you're OK, but that so-called Durant Special of yours is disqualified from competition in any and all AAA-sanctioned events."

I was wild. I had a combination that had it over everything else. From San Francisco I sent the Board a telegram of about 5,000 words, explaining all the details and asking why I should be penalized. They would not unbend. I believe to this day that if it hadn't been for that I would have been Indianapolis' first three-time winner. I had won it the year before, was about to win it the following year, and had the best car in the country at that moment.

Cliff felt badly about the whole situation and offered to buy the engine from me for $6,000. I told him that he could get one from Miller for $4,000 but he said he wouldn't give Harry $4 for an engine, that he wanted this one that was running so well, that had had so much development poured into it. So I took his generous offer and gave him back his car, with the engine in it. And there was an interesting gimmick involved. *His name* was Durant and no one could keep him from calling his own car by his own name. It was reinstated, he and Dave Lewis drove it in the "500," but only finished twelfth.

I, in the meantime, rushed out and hired everybody in Southern California to build a new car around a new Miller engine. We took it down to the express car in a wheelbarrow and put it together at the Speedway. The gas tank came loose, both of Miller's steering knuckles broke and Murphy and Hartz, in their Miller-engined Duesenbergs, finished first and second.

I have often made the remark that I am the guy who made Harry Miller. In some sense that is true, but closer inspection suggests that I was not more than an intermediary. Actually, I think it can be said in truth that Col. Hall—quite inadvertently of course, and through the agency of Duesenberg—was the guy who made Harry Miller. Hartz and Murphy were pretty quick in jumping on the bandwagon once I got my rig running, and one should not discount the fact that first-rate chauffeurs are of the utmost importance. Later, of course, others of similar caliber got aboard and Mr. Harry Miller was off and running.

chapter 19

MILLER: THE YEARS OF GLORY-I

LEO GOOSSEN UNEQUIVOCALLY CONFIRMED that Tommy Milton and Ira Vail put up the funds for the design and construction of the original Miller straight eight and added that Milton brought a Duesey eight to the Miller plant to be studied, copied, and improved upon. Memories conflict on scattered details, but on none of major importance. Just which intellects made the most significant contributions never will be known at this late date, but all the surviving participants who were interviewed decades later readily acknowledged that the Miller 183 was a synthesis of all the most advanced ideas of which they were aware. At that point in history this boiled down very simply to Peugeot, Ballot, and Duesenberg, and to using their best features as a start toward outdoing them. True to usual experience, the first prototypes were not immediately successful and required considerable development before they began to reveal their potential. But this potential was shouted to the world by Murphy's 1922 Indianapolis victory, and the blood line of purebred Miller engines was launched on its immortal way.

The race was the runaway that Milton had foreseen, but it was his one-time friend and present bitter rival who cashed in on the new power plant. Murphy was the fastest of the twenty-seven qualifiers at 100.50 MPH. Hartz was second at 99.90. In the race itself, Murphy led almost from start to finish and won 155 of the lap prizes in addition to the 20,000-dollar winner's purse, resulting in a total of 28,075 dollars. Hartz finished second but never got within two laps of the winner. Murphy's average

The winning combination of the 1923 "500." Reinforcing rods under frame were soon done away with. Early Millers used gun-metal blueing for many parts—only on radiator shell here. IMS

for the 500 miles was 94.48 MPH, devastating the existing record of 89.84 which de Palma had set in 1915 with a 274 cubic-inch, aircraft-engined Mercedes.

An important element in Murphy's victory was his mechanic, Ernie Olson. Olson gave the machine its faultless preparation, rode beside the driver throughout the race, and managed the car's four pit stops in record time. During one, a wheel was changed and fuel taken on in 38 seconds. Knowing the beating he was in for, Olson had bound his whole torso in heavy surgical gauze before the race. During its duration of 5 hours 17 minutes 31 seconds he lost eleven pounds of weight. But seeing the checkered flag there and at Le Mans were always, for Olson, the two greatest moments of his life.

Actually, the Miller-engined Murphy Special had passed its first test three weeks before when it won the 100-miler on the Cotati boards at 114.2 MPH. After Indianapolis, the Murphy-Olson team won the 300-miler at Uniontown, then repeated the feat in a 300-miler at Tacoma. This gave Murphy the 1922 AAA National Championship.

Whatever Cliff Durant's feelings about Miller were before Indianapolis that year, they underwent a thorough change as the young millionaire read the writing on the wall. He had recently resigned as vice president of the Chevrolet Motor Company of California in order to assume the management on the West Coast of his father's newly formed Durant Motors Incorporated. He still found time to organize and run the most ambitious racing team in American history.

Cliff went to Miller after the "500", and although the 183-cubic-inch formula was all but finished, he commissioned the construction of not one but six complete new race cars. He signed up much of the cream of the nation's driving talent—Dario Resta, Art Klein, Earl Cooper, Eddie Hearne, Murphy as team captain—and reserved one car for himself. The vivid yellow Durants, with their red frames and wheels, were completed in time for the big, closing event of the season, the Thanksgiving Day 250-miler on the Beverly Hills boards. Murphy's mount, bearing the numeral "1" of the National Champion, again ran away with the race, this time with an average of 114.6 MPH. The year before, Hearne's winning speed on a Duesenberg had been 109.7.

The six Durant Specials filled out the top ten in that climactic race of the season and, technically, of the formula. Cliff himself added more luster to the name of his father's new passenger car by setting a series of new absolute records in AAA-timed runs at Beverly Hills: 117 MPH for 25 miles, 118 for 50 miles, and 115 for 75 miles.

It is interesting to note that, for the 1922 season, the AAA made optional the use of one-man bodies and the elimination of the riding mechanic. At Indianapolis only one entry took advantage of this opportunity for weight-saving and streamlining. The following year every entry did so with the exception of the Mercedes, which had been designed for road racing.

Spurred by massive success, Durant placed an order with Miller for a team of new cars for the 1923 through 1925 one-man, two-liter, 122-cubic-inch formula. This, however, did not go into effect until the Indianapolis "500" at the end of May, and he got further mileage out of his 183's in the opening races of the new year. Murphy won the 250-mile opener at Beverly Hills at 115.8 MPH, then the 150-miler on the smaller Fresno board track with an average of 103.0.

Miller suddenly found himself in the unique position of building race cars on a series-production, over-the-counter basis. Instead of encouraging him to cheapen his product it gave him the wherewithal to indulge his passions for perfection and experimentation.

Taking his lead from Durant, another production-car manufacturer bought Miller cars and attached his own product name to them. This was Harry C. Stutz who, having sold the use of his name to the firm which bore it, was introducing a fine sports-touring car under the initials HCS. He bought a brace of Miller 122's just days before the race, entered them as HCS Specials, and signed Tommy Milton and Howdy Wilcox to drive them. His team was completely outnumbered by that of Durant who, between his all-new Miller 122's and converted 183's, brought eight Durant Specials to the Indianapolis starting grid. This was capacity production for Miller, but he did find time to complete one entry for himself. It was the first Miller Special in its own right and was driven by Bennett Hill.

Narrow frontal aspect of the 122. Murphy peers around cowl; Duray watches work in background. TED WILSON

Below: Two other views of the lean, functional, and graceful 122. TED WILSON

Miller's competition that year promised to include three formidable Duesenbergs. Only the Phil Shafer-Wade Morton car showed up in time to qualify, and on the day before the race at that. But the top-secret entries of the Packard Motor Car Company were on hand with a stellar team of drivers: Dario Resta, Joe Boyer, and their designer Ralph de Palma. Los Angeles by this time had become the Modena of the New World, and the cars were built at the Earl C. Anthony Packard agency there. The published rumor that the construction of many of their components had been farmed out to Miller had no basis in fact. The town had begun to bristle with racing-oriented artisans.

Another major threat was the three-car Mercedes factory team, staffed by giants Lautenschlager, Werner, and Sailer. These "ex-enemies" were barred from the all-important contests of France, but one of these cars had won the 1922 Targa Florio and such competition was welcomed by the Speedway.

Then there was the Fronty-Ford Barber-Warnock, which deserved to be taken much more seriously than it was before the race. And, finally, taken more seriously than it deserved to be, there was the Bugatti threat—five mechanical jewels, all modified Type 30's, driven by Louis Zborowski, Pierre de Vizcaya, Prince de Cystria, M. Alzaga, and Raoul Riganti, with none other than Lieutenant Colonel (Ret.) George Robertson as team manager. This assemblage was touted as a fearsome French factory entry whereas in actuality it was merely a group of wealthy amateurs which Bugatti's Paris agent had encouraged to undertake a diverting transatlantic adventure. All of these straight eight Bugattis, interestingly enough, were equipped with hydraulic brakes at the front and cable-operated brakes at the rear. The Europeans had a distinct advantage in this race since, while it was the first for two-liter cars in the United States, the formula had been in effect on the other side of the ocean for a year.

From the beginning of practice at Indianapolis in 1923 it was evident that the Americans had lavished all their effort on producing good engines for the new formula. But their chassis all were merely scaled-down versions of the much heavier 183's and, due to their lightness, they threatened to jolt themselves and their drivers to bits over the rough brick surface. The ordeal was so grueling that, of the twenty-four starters, only three drivers managed to go the entire distance without relief: Murphy and Hartz on Durant Millers, and Corum on the Barber-Warnock–Fronty-Ford.

It was a wonderful, many-sided duel and one of the most interestingly contested of all Indianapolis races. Above all, it was a bitter and long seesaw battle between arch-rivals Milton and Murphy. It was also a duel between the Durant and HCS teams. And it was a duel between America and Europe and between Germany and France. The highly modified one-man Bugattis ran extremely well, staying consistently among the front contenders. The Mercedes ran even better, and both the Germans and the French lay among the top handful of leaders for scores of laps. But one by one the Bugattis ate up their connecting-rod roller bearings, and the Mercedes drivers succumbed to fatigue and lack of experience with this very special type of racing. The Packards, the most feared of prerace threats, quickly fell by the way, two with blown head gaskets and one with a shattered final drive. The lone Duesenberg ground on dully and the surprising little Barber-Warnock stayed glued to the vanguard of the pack.

Meanwhile, HCS and Durant diced for the big stakes. Milton held the lead for the lion's share of the distance but he was constantly challenged, pressed, and passed by Murphy and often by Hartz. Cliff Durant himself drove one of his best races ever and often took the lead. At Lap 100 Milton came in for tires and handed over to Wilcox, whose own HCS had lost its clutch at Lap 60. Hartz surged into the lead, Wilcox recovered it, handed over again to Milton, who blasted home the victor at 92.44 MPH. In addition to the 20,000-dollar purse he won 9,600 dollars in lap money. He was the first driver to win Indianapolis twice.

The most fortunate of the exciting, supercharged Mercedes finished an unprofitable eighth, driven by Carl Sailer. Bugatti's best performance was ninth, thanks to the consistent driving of novice Prince de Cystria. This was Europe's last transatlantic gasp for decades except for Fiat's one-car effort in 1925. The Americans had done more than merely invent oval-track racing. Whatever their chassis may have lacked, they had learned to build engines supremely well.

Automotive Industries commented:

> The big figure is Harry Miller, the Los Angeles race car specialist, the first man who has ever properly merchandised the race car. Miller entered eleven cars in this race and six of them finished. Twenty-four cars started, and only eleven finished. This makes a very good record for Miller.

This was the barest beginning of the Miller record. On July 4 Hearne won the Kansas City 250-miler at 105.8 MPH. Two months later he won the Altoona 200-mile inaugural meet at 111.5 with his Durant Special; Wilcox crashed fatally in his HCS. On October 1 Hartz won the Fresno 150-miler and on November 29, Benny Hill averaged 112.0 MPH for 250 miles at Beverly Hills. Tommy Milton rounded out the year by showing what a Miller car could do on the dirt. He won the Syracuse 100-miler and set a new mile dirt-track record of 85.1 MPH with the HCS with which he had won Indianapolis.

Came 1924 and with it refinement of the defects that the first year of the two-liter formula had revealed. This was routine, but the year brought a pair of historic changes. One, Murphy had acquired the services of practical engineer Riley Brett, also of Los Angeles. Brett foresaw great inherent merit in front-wheel drive and convinced Murphy that a properly designed FD car could sweep everything before it. Murphy underwrote the project, and Brett worked with Miller and Goossen on its realization. Two, Duesenberg unveiled another secret weapon on which he had been laboring for years: centrifugal supercharging. Each of these rather earth-shaking innovations demands its own chapter.

The 1924 season got off to a memorable start in April with AAA straightaway record runs by Milton at Muroc Dry Lake in California. He turned two-way averages over the flying-start mile at 141.2 MPH with a Miller 122 and 151.3 with one of the old

Milton aboard his 122. It was a pure projectile. *Autocar,* LONDON

183's. Murphy was in peak form on the boards and won the June 14 25-miler at Altoona with a 114.7 MPH average. Then on July 4 he won the race for the same distance at Kansas City. Back at Altoona on September 3 he won yet another 250-miler—a joyless day due to the death of Murphy's old stablemate, Joe Boyer. Then, two weeks later Murphy faced the starter's flag at Syracuse. He had always avoided dirt tracks, not liking their unpredictable surface changes, but this was the safest dirt track in the country, and the points for the 150-mile event would clinch the championship for him again beyond any shadow of doubt.

With only twelve laps to go, Murphy powered his way around the first turn, battling to pass the eventual winner, Phil Shafer. He was well crossed up, and when he corrected his steering to enter the backstretch he was moving so fast that his Miller followed its original trajectory. He hit the inside rail, bounced off it, careened into it again, bounced again, and then crashed through the railing. It impaled him.

But before this tragedy there was the 1924 "500," one of the most telling contests in American racing history. Now the 122's were highly perfect both in engine and chassis, as was proved by only five out of the twenty-two starters failing to complete the distance. It was proved too by the speeds of the cars: the averages of the first five finishers all exceeded the old all-time record of 94.48 MPH. The victory of course went to Duesenberg and his vindicated faith in the centrifugal supercharger. But, as *Motor Age* commented:

> The victory was not a walkaway. It was the bitterly contested, fluctuating, tense, dogged, persistent, thrilling thing that the lovers of automobile racing like to see, with fortune frowning and then smiling upon one or another of the little demons of velocity.

Among the twenty-two qualifiers were fourteen Millers, including five Durant Specials and Earl Cooper's Studebaker Special which was to win the Fresno 150-miler later in the year. There were just four Duesenbergs, three of which were supercharged. Three Barber-Warnock Fronties and a Schmidt Special (actually Lautenschlager's former Mercedes which had been purchased from Louis Chevrolet) completed the field.

In qualifying, Miller's terrible trio—Murphy, Milton, and Hartz—led the field with four-lap averages of 108, 105, and 107 MPH. With the exception of Duesenbergs in fourth and tenth spots, Miller cars filled the top ten, Boyer's Duesey and Benny Hill's Miller tying for fourth place. *Motor Age* continued:

> It was a start such as Indianapolis had never before witnessed and the stands went wild with the joy of it—and the whisper passed, "It's the supercharger." And it was the supercharger.

Boyer flashed into the lead on the first lap, leaving all the field hopelessly in his wake. But before the lap was finished a key sheared in his blower drive, and Number 9 Duesey fell to the back of the pack. Now it became a Miller parade, with the brilliant Murphy in the lead but forced to trade it constantly with Cooper's green Studebaker Miller. Hill lay a consistent third while Corum's Number 15 Duesenberg maintained

on Lap 24, then relieved Ben Shoaf. Stapp, in Shoaf's Perfect Circle Duesenberg Special, had second place secured when, just five miles from the finish, his rear axle gears failed. The nearest pursuer was ten miles to the rear. The Shoaf machine was, along with Wade Morton's Thompson Valve Special, one of the new radically offset Duesey chassis with ring and pinion gears located adjacent to the left rear wheel. The gears had to be specially machined to accommodate the diagonal drive shaft, and they turned out to be the weak link in the system. Morton had been relieved by Fred Winnai when, on Lap 152 and perhaps also due to failure of the unusual rear axle gears, the car hit the wall and burned. Duesenberg's best showing in the "500" was Dave Evans' fifth place in a car with a conventional rear axle. The sixth place credited to de Paolo and Duesenberg at the time was in error; Pete's car was a Miller front-drive. However, he did give the marque its only other win of the season with an average of 116.6 MPH for 200 miles at Altoona on June 11.

In 1928 Duesenberg's luck was equal or worse. The cars of Slim Corum and Dutch Baumann were wrecked in practice at Indianapolis. Leon Duray qualified his Miller front-drive at 122.391, but the fastest Duesey was Jimmy Gleason's at only 111.708. In the race both Benny Shoaf and Ira Hall were doing well but both were eliminated by stupid minor collisions which were not their fault. Wrote Jerry Gebby in the *Auburn-Cord-Duesenberg Newsletter:*

> But the real catastrophe occurred when Jimmy Gleason stopped at his pit on Lap 195, only twelve and a half miles from the finish. He had held a nice lead in first place for the past forty laps, but the engine was dead when he coasted in. Steam was pouring from under the hood and the cause was found to be a water hose that had come off the manifold, allowing water to run onto the spark plugs and distributor. While the engine was being dried, re-filled and started again, the entire money field went past the unfortunate Gleason, plus two more cars. He left the pit in thirteenth place and was unable to make up any of his loss in the remaining five laps. Twenty thousand dollars went down the drain with that loose hose, but automobile racing has always been this way.

Fred Frame finished eighth in that race in a privately owned Duesey. The rest of the year was a washout except for Winnai's slow 101 MPH win in a 100-miler at Atlantic City.

In 1929 seven Duesies were entered in the "500." Bob Robinson's was not finished in time to attempt to qualify. Thane Houser's supercharger drive broke with no time left for repair. Frank Swigert's 99.585 MPH was too slow to get him in the race. Ernie Triplett had the fastest qualifying time for a Duesenberg at 114.789, which was none too promising in view of the many faster cars in the field, up to and including Cliff Woodbury's Miller at 120.599. The other Duesies to make the race were those of Gleason, Winnai, and Bill Spence.

The race got off to a bad start for all concerned, particularly for Duesenberg, when on Lap 9 Spence skidded into the wall at high speed in the southeast turn. The car overturned and slid, crushing the driver fatally. On Lap 47 Triplett left the fray with a broken connecting rod. But Gleason finished third, less than four MPH slower than the winning Miller's 97.585. His performance was particularly admirable since his car was a 1923 machine whose 122-cubic-inch engine had been replaced with a 91 which was equipped with one of the early superchargers. Winnai, in a newer car with one of the latest blowers, finished fifth. Elsewhere on the Championship Trail that season the best performance by the marque was Winnai's third in the Syracuse 100-miler.

In 1930 the old order changed and the 366-cubic-inch formula was introduced. Fred and Augie, neither of them with any great hopes for their racing future, decided to take different paths. Fred chose to do his racing with modified Duesenberg Model A passenger-car components, while Augie remained faithful to the wiry, small-displacement thoroughbreds.

Between them and several private owners nine Duesenberg-powered cars were qualified for the 1930 "500." Augie's cars for Dave Evans, Deacon Litz, and Babe Stapp had displacements of 138, 150, and 142.5 cubic inches respectively.

In Fred's camp William Denver and Rickliffe Decker showed up with two of the 1923 Indianapolis Mercedes, which were later owned by Louis Chevrolet. According to *Motor Age* these two cars were fitted with the 300-cubic-inch engines from Tommy Milton's Land Speed Record machine. If this was indeed true the three-valve-per-cylinder, 1919 engines did not qualify under the new rules and were replaced with slightly modified Model A Duesey engines. The Model A-powered machines were the fastest qualifiers in the semi-stock class, with Bill Cummings averaging 106.173 MPH for his ten miles. The quickest of Augie's enlarged 91's was qualified by Deacon Litz at 105.755. But no Duesey was even close to the eventual winner—Billy Arnold in a Hartz-Miller front-drive which qualified at 113.263. Still, anything can happen in a race, and Duesenberg chances were good. But everything happened to Duesenberg, almost from the moment that Wade Morton pulled his Cord pace car off the course at 80 MPH. *Motor* reported:

> On the very first lap, on the first turn, not half a mile from the starting line, Chet Gardner in a Duesenberg took a skid which might well have tangled up at least half the thirty-eight cars in what might have been the greatest catastrophe in automobile racing history. . . .
>
> As it was, a spill occurred early in the race which involved seven cars, while others evaded the mixup by a hair's breadth. Red Roberts, on his 20th lap, went into a spin entering the north turn (after leaving the back stretch). Peter de Paolo had relinquished the wheel of this car [Duesenberg] on the 8th lap because he said it was hard to steer. Stapp [Duesenberg] hit Roberts. Litz [Duesenberg] hit Trexler. Lou Moore climbed the wall. Johnny Seymour's left front hub momentarily locked into the wires on Jimmy Gleason's right rear wheel and later Seymour's car hit the wall. With wheel locked, Gleason stepped on the gas and as a result broke a timing gear and bent some valves.

This early elimination of three cars was only the beginning of disaster for the

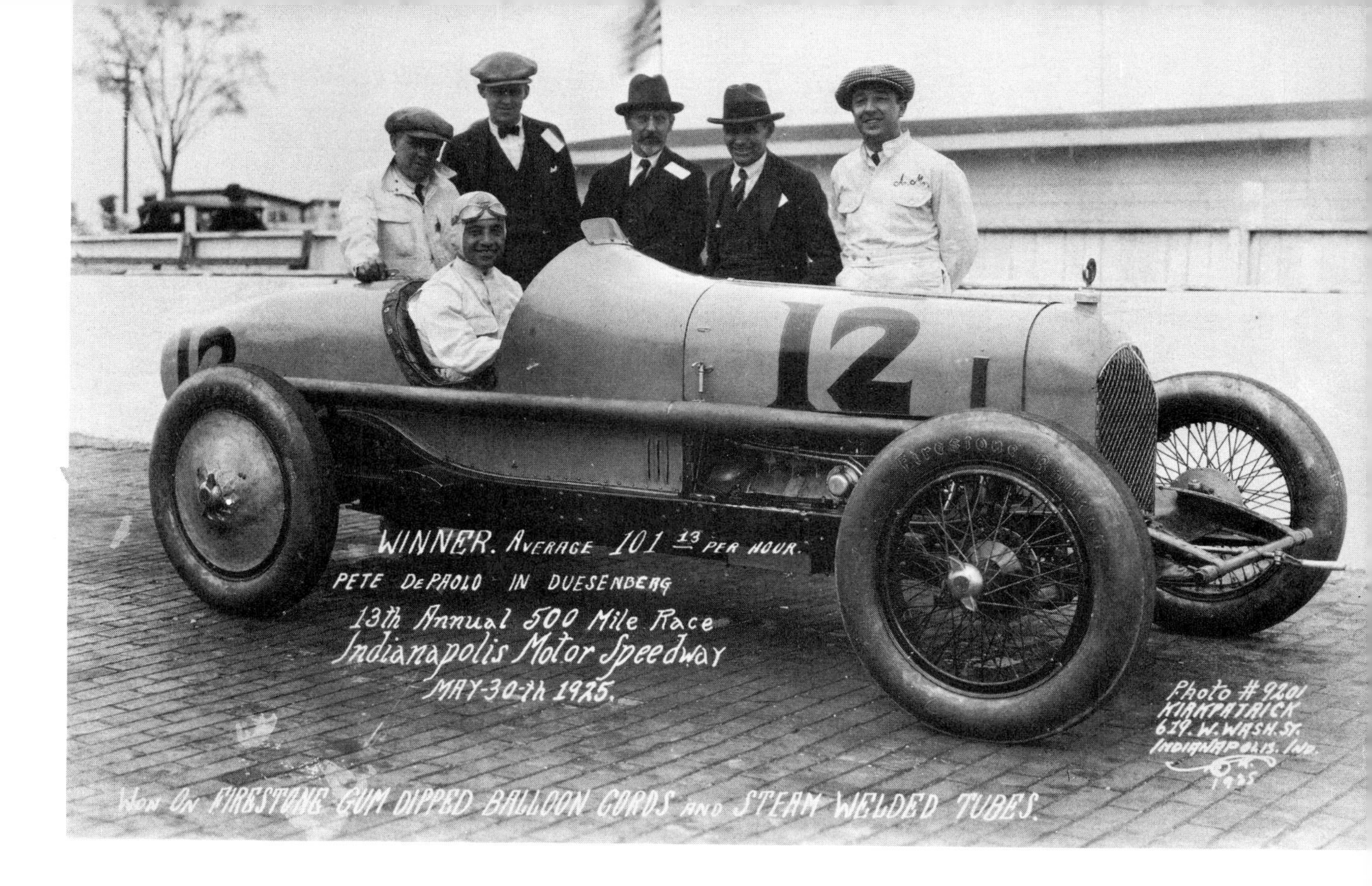

In winning the 1925 "500" Pete de Paolo was the first driver ever to exceed 100 MPH for the distance. Supercharger expert Dr. Sanford Moss stands below flag. IMS

Duesies. On Lap 29 Cy Marshall in one of the Model A's spun out on the north turn and crashed through the retaining wall. His riding mechanic, his brother Charles, died on the spot, and Cy was gravely injured. Rick Decker, driving one of the Mercedes Model A's, skidded on Lap 42 and smacked the wall with his right rear wheel. He came into the pits, spent four minutes checking for possible damage, then took off again. But on the next lap the car's torque-tube yoke fractured and it, too, was eliminated. Almost simultaneously Bill Denver's sister car threw a connecting rod.

This left only Cummings' Model A and Evans' 138-cubic-inch "91" to defend the Duesey name. They did a beautiful job, Cummings finishing fifth in his rookie year with the fastest of the semi-stocks at 93.579 MPH. Evans came in immediately behind him with an average of 92.571 for the 500 miles. The only Duesenberg victories in the Big Time that year were Cummings' in the opening race of the season, a 100-miler at Langhorne, and in the final event at Syracuse, for the same distance. Both were one-mile dirt tracks.

Duesenberg's years of glory expired with the 91.5-cubic-inch formula. Fred Frame did finish a splended second at Indianapolis in 1931 in one of Augie's enlarged 91's—prepared, significantly, by wizard Harry Hartz. After that, only bigger engines of new design could hope to be competitive, and they were not forthcoming from the house of the Big D, the Winged Eight.

chapter 13

DUESENBERG: TOMMY MILTON AND THE LAND SPEED RECORD

EACH NEW ATTACK on the record for ultimate speed is the horizontal equivalent of an assault on an unclimbed Everest. Each new attempt is a voyage into the perilous unknown, and each one takes much of the courage that a man possesses. Camille Jenatzy's 65.79 MPH absolute record was just as much a death-defying feat in 1899 as Bob Summers' 409.695 with piston engines or Craig Breedlove's jet-propelled 600-plus in 1965. Again, the world always wonders, "Why do they do it?" and again the answer goes much deeper than the hope of material reward; again it is rooted in the passion and challenge of the machine and in proving one's self to one's self and to the world. All that Tommy Milton ever got for pushing the Land Speed Record to 156.046 MPH in 1920 was a sterling silver tea service from the Goodyear Tire & Rubber Company.

At Uniontown on September 1, 1919, the Great Milton (only one other driver, Earl Cooper, ever amassed more Championship points in his career) seemed to have the 225-mile race won hands down. He was a full lap ahead of the second-place Gaston Chevrolet—Joe Boyer Frontenac as he flew down the grandstand straightaway at well over 100 MPH. Then a great gasp went up from the crowd of 40,000 as a cloud of greasy black smoke erupted from beneath the Duesenberg's hood, and sheets of raw, red flame enveloped its cockpit. Milton reacted automatically. With one hand he

After the successful record runs on the Daytona sands. Bare spot on hood is a result of engine-compartment fire. BILL TUTHILL

warped the front wheels over to full lock, and with the other he yanked the hand brake and locked the rear wheels. The brake lever was a stout steel bar but it bent like a lead pipe under Milton's madly adrenalized grip. The car pivoted on the proverbial dime, rolled backwards as the wind blew the flames away from the driver and riding mechanic Dwight Kessler, and came to a stop. The gravely burned pair were pulled from the car and rushed to the local hospital. Milton refused to authorize the doctors to amputate his leg, spent nine weeks there on his back, then left on crutches for his great record-breaking rendezvous with the 16-valve aero engine at Sheepshead Bay. Before that a little history was made in Milton's hospital room.

Jimmy Murphy came to visit him, in fact, to say good-by. Murphy, who happened to be an orphan, was a very complex young man: wistful, gentle, disarming, totally charming, and, beneath all this, aggressive as hell. Milton liked and befriended him from the day the twenty-three-year-old Murphy took a mechanic's job with Duesenberg early in 1919. "It takes one to know one," and Milton recognized Murphy's class, made him his riding mechanic, and agreed that he had the stuff of which good drivers are made.

At the Uniontown Hospital Murphy unburdened himself to his mentor. It had become obvious that Fred had no intention of granting his ambition, so Jimmy was going back to his old garage-mechanic job in Vernon, California. Milton urged him to hang on until the start of the 1920 season. In the meantime he, Milton, the captain of the Duesenberg team, would lay down the law to Fred that either Milton's protégé was given a car to drive or he, Milton, would quit. Murphy stayed. Milton was good to his promise. Murphy got his ride and more than vindicated Milton's faith in his ability.

While Milton lay on his back for better than two months he had nothing but time in which to plan and dream, and it was then that he hatched his plan to become the fastest man on earth. The 300-cubic-inch formula was dead, and Fred had three big straight-eight engines which now were useless. Assuming correctly that he could make use of them for the good of the cause, Milton whiled away the long days designing a 16-cylinder, 600-cubic-inch record machine. When he communicated the plan to Fred, along with his willingness to shoulder its cost, he was told to fire up his crutches and hustle to Elizabeth.

I had many wonderful sessions with Tommy Milton, and we corresponded for years. At one point I praised his unique and vivid style and he replied:

> I am a bit unhappy to learn that I wear my ego so conspicuously. As a matter of fact I do like language and, who knows, but one day I shall be so brash as to take a hand at an auto-biography. Inasmuch as I have been thinking about this for some twenty-five years it seems a reasonable presumption that it will never occur.

This is the reason for the liberal use of long quotes in this book. Nothing is easier than to paraphrase source material. It makes the author seem authoritative but at the cost of true, first-hand authority. There is no substitute for authentic experience nor for the vitality of language such as Milton's.

> In the construction of this car I was one of the busiest little beavers you ever heard of. I sawed out the frame from flat stock and swung a 16-pound sledge for the blacksmith who formed the rails. The power plant consisted of two straight-eight competition engines which were connected directly to a common solid rear axle which had two pinions and two ring gears. Fortunately, we were then using cone clutches. I say fortunately because the Contest Board had told us that no reverse gear would be required. When we had arrived at Daytona there was a reversal of this opinion which I need hardly tell you posed a pretty serious problem. We hurdled this by attaching a flywheel starter gear to the cone of one of the clutches and by making a leather wheel which, with benefit of Rube Goldberg linkage, we were able to press into the crotch of the two flywheels. Since the arrangement was the plan of Fred and Augie Duesenberg, I need feel no embarrassment in stating that it was by any yardstick an ingenious solution. The thing actually worked and movies were taken of the car going backward on the beach.
>
> The solid rear axle was the first one I had ever seen that was live. In other words, the axle shafts themselves supported the weight and were not contained

in the usual conventional housing. The bearings were carried in the outer ends of the differential casings. Many competition cars have subsequently been built with this type of axle, but this may have been the first one.

Evidently the 300-cubic-inch engines had been stored in Los Angeles because when the team went to contest the Beverly Hills 250-miler (Murphy's first victory) on February 29, 1920, the beach-car chassis was taken along, and the engines were installed there.

Due to my Uniontown crash and the investment in the beach car my funds were depleted and so I welcomed the opportunity to pick up a few dollars racing at Tropical Park in Havana. I sent Murphy and Harry Hartz with the beach car to Daytona to get it ready for the record attempt.

When I got back to Key West and picked up the morning paper it said in screaming headlines that Murphy had smashed de Palma's world record of 149.9 MPH with a speed of 152—in my car, which I had paid for, had actually built and was paying him to be there preparing. I don't think that the world ever has looked so black before or since. I could have killed him when I got there; he understood that and got out.

Milton slaved over the car and ran and ran but could not get up to Murphy's speed. The gossip grew that Tom didn't have the guts to get his foot into the carburetor as far as Jimmy had. So he finally decided that his only hope was to tear the engine down to the last lock-washer, rid them of the sand that they had inhaled, correct all their clearances, and make a final try. The car's shelter was an open shed on the beach through which sand blew continuously. Milton rented enough tarpaulins to erect a tight enclosure, and he and Hartz did the job. If this effort didn't help there was nothing to do but pack up and go home in defeat. During the rebuild Milton got a steel sliver in his one good eye, which hardly helped. Finally the car was as ready as it ever would be.

The men of Duesenberg never had enough money to do things in a first-class manner and the crew at Daytona was a mere handful. The town offered no aid in the policing of the course and the various approaches had to be guarded by whatever voluntary manpower could be scrounged. Milton waited at the south end of the course until he felt that all hands had had ample time to reach their guard-posts on the approach roads to the beach, and then he charged off on his first official run. As he headed, full-bore, under a pier that crossed the course, a citizen in a Model T Ford touring car pulled up precisely in the path that Milton was committed to follow. Seeing the projectile hurtling toward him, the wandering motorist seemed to panic, and Milton watched him futilely jumping up and down and waving his arms. Milton made the slightest of swerves, which left the Model T half-buried in sand, and kept his foot all the way down until he crossed the finish line. Back to Milton:

The car had its faults. One was that the exhaust pipe from the left-hand engine ran through the cockpit. The heat was pretty terrific. Then, the centrally mounted steering column was an open tube. One of the engines had its crankcase breathers in the center and when I got under way oil vapors came streaming up through the hollow tube, covered the windscreen and my goggles, with the result that visibility became practically zero, zero. Actually I finished that northbound run by using the spectators on the beach as a landmark.

The return run went well enough until smoke, heat and fumes announced that we were on fire; I was carrying a full-length underpan in which there was undoubtedly oil and gasoline to support the blaze that cooked the paint off the hood. I took a squint at the aluminum firewall, which seemed not to be melting. The zero station was not too distant and it was, of course, highly desirable to finish the mile since I had no way of knowing how serious the fire damage might

Right: Milton with record machine in exhibition appearance in Indianapolis. Note how upper exhaust pipe passes through cockpit. IMS

Below: Power source of the LSR Duesenberg was a pair of straight-eight, 300-cubic-inch SOHC racing engines. TOMMY MILTON

Note hollow steering column, path of exhaust pipe from left-hand engine, reverse-gear device between and above flywheels, direct drive to two final-drive units. TOMMY MILTON

be. Having finished the mile I got close to the water's edge with the idea of dunking her but this did not prove necessary. It was fortunate that we were able to complete the run because there was considerable damage—probably two or three days' work—and the season when good sand can be expected already had passed. Extinguishing the blaze was laborious because the hood was bolted on; when finally removed the fire was put out with beach sand. Having been rather severely burned at Uniontown just a few months before, this experience was certainly unwelcome, if not alarming. But it did give me the satisfaction of beating both de Palma *and* Murphy.

Although this resulted in practically a total breach between Murphy and myself, the mellowing influence of time impels me to say that there were extenuating circumstances. Those who knew Jimmy will attest to the remarkable personality which was his, and which influenced Fred Duesenberg in particular. I know beyond any peradventure of doubt that Fred urged Jimmy to run the car all-out. Furthermore it is—or at least it was then—the fervent ambition of any youngster to become the World's Speed King. Undoubtedly the temptation was tremendous to the extent that, momentarily at least, it overrode his evaluation of loyalty. At least in retrospect it was purely a personal feud and I have always been happy that the conflict did not get a public airing. The final word on the matter is that after his untimely death at Syracuse in 1924 I was privileged to escort the body to Los Angeles.

Milton was modest. Actually it was he who came forward and shouldered all the responsibilities attendant upon Murphy's death. As for Fred Duesenberg, he was finished with the man. He was committed to drive at Indianapolis a month after Daytona. He finished an excellent third, after Gaston Chevrolet's Monroe and René Thomas' Ballot, then resigned from the team.

chapter 14

DUESENBERG: WHEN AMERICA WON THE FRENCH GRAND-PRIX

FOLLOWING THE THOMAS FLYER'S WIN in the strange New York to Paris "race" of 1908, via North America and Asia, American contestants wisely stayed on their own side of the water. After the first World War, however, it quickly became clear that the quality of American racing machinery had been radically transformed.

Some powers in France made it known that American cars would be warmly welcomed as entries for the 1921 Grand Prix of France, Europe's and the world's most important race. But others pushed through regulations that were designed to make this a one-make race, meaning no race at all. Instead of preliminary time trials for the weeding out of final entries, dynamometer tests were decreed. It was bad enough that hardly anyone had a dyno, least of all the Automobile Club of France, but it also was required that the three-liter engines should develop no less than 30 BHP at 1,000 RPM and 90 at 3,000 RPM. This, in effect would give Ballot the victory without his ever having to bring his cars to the starting line. Said *Automotive Industries:*

> Owing to the general dissatisfaction with the regulations the outlook for the 1921 Grand Prix is anything but bright. There is very little chance for a team from this country competing, because none of our large manufacturers participates even in our big classic, and to most of those who have been building racing cars

Starting lineup of the 1921 French Grand Prix. Murphy is Number 12. GB COLLECTION

in America during the past several years the expense of a Grand Prix venture would undoubtedly be prohibitive.

This was a very chaotic period, made more so by the fact that the Automobile Club of France labored under the impression that racing in the United States was controlled by the small New York organization which was called the Automobile Club of America. Finally, with the assistance of the ACA, the ACF was persuaded that the Contest Board of the AAA was the national governing body, and meaningful dialogues could begin. The end result was that the ACF, through Charles Faroux, advised the AAA that it had adopted the current Indianapolis rules in their entirety and, to make things more attractive, planned to stage its Grand Prix in a part of France where American popularity was at its highest. The bench-test requirement was abandoned and the tricolor carpet rolled out. Still there were no Stateside takers.

The situation was nearly as grim in France. Ballot had a strong four-car team and a magnificent stable of drivers: Louis Wagner, Jean Chassagne, Jules Goux, and Ralph de Palma. Fiat announced that it would enter but in the same breath let insiders know that it had no intention of entering. Sunbeam-Talbot-Darracq announced its participation with no less than eight cars. But then chief engineer Louis Coatalen chose this time to go off to relax on the Isle of Capri, the various shop heads fell to

Olson, Augie, and Murphy, setting sail for France; Murphy's hand had been burned in a minor accident. GB COLLECTION

engaging in knock-down, drag-out physical combat, and the cars did not get built. At the eleventh hour two British office employees took the situation into their own hands, whipped the troops into line, worked with the tools themselves night and day, and got some cars ready. Then there was a single 1.5-liter Mathis entered for the ride, and that was all.

Of course American passenger-car manufacturers had nothing that they could pit against even this modest field. But Fred Duesenberg thought that he might have something, and it tied in with his own new passenger-car program. One fine day he shot a cable to W. F. Bradley: "ENTER THREE DUESENBERGS IN GP." He neglected to send entry money.

Bradley knew full well that the Duesenberg name was one to be reckoned with. He went directly to the ACF's palace in the Place de la Concorde, where it still is today, and encountered Baron René de Knyff in front of its massive portals.

"I have three good American cars for your race," he said. "The entry fee hasn't come through yet, but if you'll give me just a few days I know you'll have it."

De Knyff, a compulsive stickler for form, cast his eyes up at the palace and said, "Why don't you just ask me to set fire to this building?"

Bradley was doing the man a huge favor, bringing him cars to fill out a race that very likely never would start. He felt that he was entitled to this small courtesy in return, but it was haughtily refused.

The deadline was at hand, and Bradley cabled Duesenberg to this effect, advising him that there still was a month's period of grace during which entries could be made, but at double the fat entry fee. The man who paid it and for a fourth entry to boot and for the transportation of the team and the cars was none other than Albert Champion. His was a Franco-American rags-to-riches success story, and it pleased him to spend a little of his American-made wealth on this cultural mission.

Meanwhile, due to their formidable internal difficulties, the STD team was withdrawn, as Bradley had known it might be, and Fiat bowed out officially. This left only the Ballots and the lone little Mathis. The ACF prepared to cancel the race the instant the grace period expired.

Bradley could be as difficult as those he had to deal with, and he held the Duesenberg entry money for a solid week. Came the final day, with the French Grand Prix promising to be a total fiasco, and Bradley still stayed away from the Place de la Concorde. De Knyff's office closed at six P.M. and Bradley waited until the first clock began to toll. The bell towers of Paris all were minutes apart and he knew what he was doing when he walked past the ancient little bird-cage elevator that still goes to the CSI offices on the top floor and headed up the stairs. The last clock was striking when he flung open the door, tossed the money on the desk, and said to the Baron, "This is for four Duesenbergs. Now perhaps you'll have a race."

That saved the first postwar French Grand Prix. Then, twenty-four hours before the start, STD announced that it would be competing after all, with four cars.

Next door to the ACF is the majestic Hotel Crillon, the old palace in which the Treaty of 1778 was signed, by which France recognized the independence of the United States. In 1921 it was the most distinguished hotel in Paris and the haunt of the entire diplomatic corps. It was there that Ralph de Palma told Ernest Ballot he wanted to stay, and his patron approved. De Palma, who still wore the aura of one of the greatest drivers of all time, was as much a natural aristocrat as he was a born showman. He was one of the most cherished guests ever hosted by the Crillon, which was in a position to choose its guests and did so. In Bradley's opinion the United States never had a more effective ambassador.

Duesenberg, as usual, arrived in cliff-hanger style, with only hours to spare but, for once, with excellent advance preparation. American race driver George Robertson, a giant of the old Vanderbilt Cup days, had spent the war years in France. He held the rank of lieutenant colonel with the Bolling Mission and spent most of the duration managing the movement of American military vehicles in and out of the important port of Bordeaux. Fred Duesenberg had secured Robertson's services as team manager

The winner at speed. ERNIE OLSON

in France, and he sent Augie over to take care of the mechanical side of the campaign as he alone could. The press agreed that while the team suffered from lack of road-racing experience its organization was far superior to that of its rivals.

The Duesenberg drivers were to be Jimmy Murphy, Joe Boyer, Albert Guyot, and Louis Inghibert. The course was laid out over 10.7 miles of public roads, consisting of stone beds over which a sand composition had been spread and rolled. It was a slippery surface, and practice was filled with incidents.

One of the Ballots crashed, was a total loss, and its driver, Renard, was killed. This entry was replaced by the only other available Ballot—a two-liter four-cylinder, which Goux drove. Murphy's brakes locked when entering a turn, his car got sideways and landed upside down in a ditch; putting him in the hospital with a broken rib. Team driver Inghibert had been riding with Murphy, and his injuries were so serious that he could not hope to drive in the race. His Duesenberg was turned over to wine magnate André Dubonnet, all his life an avid motorist, who, incidentally, gave General Motors its first independent front-suspension system and later became one of the directors of Simca.

During the days of practice French carburetor manufacturer Claudel made it financially interesting for all entries to use its products. Murphy alone among the Duesey drivers refused, preferring to stick with his tried-and-true Miller carburetor. This proved to be a wise move because during the race both Duesenbergs and Ballots had trouble with grit jamming their Claudels' rotary throttle valves. For all except Murphy time had to be taken out during the race for the fitting of supplementary throttle-return springs.

Important changes had taken place in tire technology. The straight-side tire had come into being, in opposition to the old clincher type, but it was as controversial and unproved as the balloon tire would be until Memorial Day, 1925. Ballot had exclusive use of Pirelli's new straight-sides and Duesenberg came loaded with gum-dipped Oldfields, products of Barney's association with Firestone. The STD's used conventional racing Dunlops. All seemed equally good in practice.

It has often been said that Duesenberg's great advantage in this race lay in its four-wheel brakes, but this is incorrect; *all* entries had them. The French used the Perrot system of ball-joint mechanical linkage, which Bendix was to adopt in the United States.

What Duesenberg did have that was unique was the Wagner-type hydraulic system, taken from the Model A passenger-car prototype. It was a novelty that attracted almost no pre-race attention, but its advantage of perfectly equal effort applied to each of the four brakes and therefore much more efficient braking and automatic adjustment permitted the Duesies to drive significantly deeper into the course's many turns. But the system still was full of bugs and only was made to work properly during the practice period. Ernie Olson recalled:

> At first we had too much braking on the front and every time Jimmy would hit the pedal the car would start fishtailing all over the road. Finally he landed in

the ditch and got badly banged up. We all worried about this crippling problem. As I recall we had 14-inch drums front and rear. I noted that the Ballots' four-wheel brakes were working very well and that their front drums were smaller than the rear. I went to Augie and suggested that we try the same effect by getting rid of some of our front brake lining. He told me to try it and I took a hacksaw and chopped two inches of that metallic lining from each of the front shoes. Jimmy was still in the hospital and I got Joe Boyer to take the car out and try it. We hurtled down the straight, he hit the brakes and the car just squatted. It tried to bury itself in the road without a bobble. We really had it over the French after that.

The Americans learned something else in practice. The French all carried spare tires and wheels. The Americans ran tests that proved to them that it was quicker to drive on the rim, even if a blowout occurred just after the pits, than to change wheels on the course and then stop again to pick up another spare wheel and tire. The French would not consider running without spares; it had never been done.

Among all the events of the practice days perhaps the most decisive was the final rift that occurred between Ralph de Palma and Ernest Ballot, although Ballot himself never was fully aware of it. De Palma was recognized universally as one of the world's finest and fastest drivers, which is why Ballot had hired the foreigner. It was not that Ralph was too lazy to change gears for himself. It was simply that he, too, took a highly analytical approach to his craft and had proved to his own satisfaction that he could gain perhaps a tenth of a second on every gear change if he would keep both hands on the wheel, nod to his riding mechanic, punch the clutch pedal, and let the mechanic do the shifting. Those tenths could add up and win races. De Palma was anticipating the preselector transmissions of the Thirties, and the automatic transmission revolution brought about by Chaparral in 1965.

Ralph knew the imperious Ballot well enough to know that he never would authorize such unorthodoxy. So Ralph explained to him that he shifted best with the lever on top of the gearbox, American style, rather than against the frame in the standard European location. Ballot accepted that, the change was made, and Ralph and his nephew, Pete de Paolo, went off to perfect their technique. Nod, depress clutch, Pete throw the shift, release clutch, keep charging, gain those tenths. They quickly reduced it to a science.

And then one day during practice, bolting down the straight into the Mulsanne bend—a 90-degree that required downshifting to second and perhaps to first, Pete missed the shift. The gears grated and the engine over-revved wildly. And Ernest Ballot happened to be standing right there, taking it all in. He reacted with pure fury.

Ballot had the shift lever returned to its original position. He did not care that de Palma perhaps had a point; he would not tolerate discussion. And that was the end of the romance between de Palma and Ballot.

The car was the same one which Ralph had driven at Indianapolis less than a month before. It had a fresh engine, which it sorely needed, but Ralph swore that it was five MPH slower than the old one. He drove the race, but Ballot had shown contempt for his efforts at winning strategy; and therefore why should Ralph extend himself for a victory that his patron wanted only on his own know-it-all, haughty diploma-engineer terms? Ralph's heart was no longer in the race—there was no money in it anyway—and afterwards he told friends, "Well, it cost me some glory but at least I've learned everything about Ballot and his cars that anyone needs to know."

That knowledge, fed into Packard's last racing effort—for the coming 122-cubic-inch formula—was not enough. The beautiful little Packard never raced successfully, and de Palma's career in the big time was at its end.

The historic 1921 French GP took place on July 25 with Murphy, taped from hips to armpits, ready on the grid with Olson beside him. The thirteen starters were lined up in pairs, the little Mathis sitting alone in the rear. Each pair was flagged off at thirty-second intervals, and the pace was terrific from the start. De Palma and Boyer both averaged 78 MPH for the first of the thirty laps, followed by Murphy and Chassagne, neck-and-neck. The second time around Murphy and Boyer were in the lead and running marvelously. On Lap 3 Chassagne overtook de Palma, and the leaders remained in that order until the end of Lap 6 when Boyer decided he had better stop to check his tires.

The STD cars already were in severe trouble. The surface of the course began to disintegrate by the end of the first hour; the sand flung aside, leaving only the coarse rock of the roadbed. It was like driving in a hail of shrapnel, rocks the size of fists flying at closing velocities of far over 100 MPH. The STD tires disintegrated almost as fast as they could be changed. Their complete unsuitability was recognized immediately, and the team rushed through the pits and parking areas, buying, borrowing, begging all the tires they could find that would fit their rims. The problem was so severe that they quickly ran out of ready-mounted spares, and the STD cars had to wait idly in the pits while new tires were pried onto their wheels. These machines, though the heaviest in the field, were extremely powerful and fast, and with more rugged rubber their performance would have made for a very different story.

On Lap 7 Murphy also dashed into the pits to check his tires. Chassagne, on the Pirellis that proved to wear like iron, surged into the lead as Boyer roared back into second place. Murphy's stop lasted mere seconds, and he quickly overtook Boyer and went after the leading Ballot. But Chassagne remained in front until Lap 17, when he was forced to retire with gasoline pouring from a stone-riddled tank.

Murphy took over the lead, followed closely by Boyer, whose Duesey threw a connecting rod on Lap 18. Then Guyot's Duesenberg moved into second, a position which he might well have held for the remaining laps. Then a great, hurtling rock cracked his riding mechanic on the skull, and when he stopped at his pit the man was too dazed to be able to crank the engine. Old veteran Arthur Duray, just a spectator in the crowd, saw what was happening, vaulted the fence, pushed the mechanic aside,

An early 122. Murphy and Miller caress it. Note damascened brake backing plates. There are three fuel tanks, two within frame rails. Oil tank is in cowl. ART STREIB

Left: Ultimate refinement: the 91. Thick plates on crankcase and bronze supercharger backing plate and gear housing mark this as a late model. GB Right: Ernie Olson and details of the 91 engine. GB COLLECTION

cubic centimeters of ethyl fluid, ran better. Use of the additive reached a proportion of 8 cubic centimeters per gallon during the 122 formula, compression ratios were increased proportionally, and output reached 122 BHP at 5,000 RPM. It was not unusual, however, for the short-stroke engines to be wound out to 6,500 RPM. Standard carburetion continued to be by four 1.5 inch, twin-throat Millers. The engine weighed 303 pounds without magneto.

Starting with the 122's built for the 1924 season it was again decided to replace the ball rear main bearing with a poured one. Up to this point chrome vanadium steel had been used for all stressed parts such as crankshafts, camshafts, and for the exquisitely made tubular connecting rods. Miller's most important change for 1924 was the substitution of nickel molybdenum alloy in these parts.

Following the Indianapolis race *Automotive Industries* observed:

> From a technical viewpoint the 1924 "500" will stand out as a high point in American racing history. Among other things it has brought home three significant facts any one of which is as important as the lessons learned from the first foreign invasion by Peugeot in 1913.
>
> First, this race has shown that only one year of development and competition is required to bring a new design to a very high state of perfection.
>
> Second, the supercharger has brought greater speed with reduced piston displacement.
>
> Third, the performance of the contenders proves that they were not only the fastest built, but also the most reliable array of race cars that ever negotiated 500 miles on the famous Indianapolis Speedway.

When Duesenberg appeared with his supercharger at Indianapolis in 1924 Miller claimed to have had a similar device under development. In 1924, however, he produced only two or three 122's that were so equipped. Without magneto they weighed only about 340 pounds and developed 236 BHP at 5,800 RPM. Between twenty-five and thirty 122's were made.

Miller and Duesenberg were friendly rivals, and before Miller had completed the development of his supercharged 122 he received a visit from Fred, who reported the following to the Society of Automotive Engineers during a meeting in the winter of 1926:

> About eight years ago we built a tractor engine for the Government. It was a four cylinder that had 1,200 cubic inches. When it was tested Mr. Miller sent Dave Lewis over to fit the carburetor. At that time, at 1,500 RPM, we got 205 horsepower. The man who was looking after the job [for the government] thought that was very remarkable. We had a two inch Miller carburetor on it that semed to act very well. While I was in Los Angeles on Thanksgiving Day in 1925 Mr. Miller gave me a test sheet of a run of a 122 cubic inch engine—just one-tenth the size of the one we had tested several years before. I believe he also used a two inch carburetor. That engine developed 203 horsepower. In eight years there has been a gain of 900 per cent. That is pretty good development.

The 122-cubic-inch formula had failed in its intended purpose, which was to create slower racing, and so was superseded in 1926 by the 1.5-liter, 91.5-cubic-inch formula. Supercharging was still too new and too experimental to be included as a handicap factor in the new regulations.

Just as the Miller 122 was an improved and scaled-down 183, so the 91 was a better, smaller, and more refined version of the 122. A major change was that while the supercharged 122 had used a blower drive which was improvised off the end of one of the camshafts, the 91 was designed from scratch as a blown engine. The blower was driven by a train of spur gears at the flywheel end of the crankshaft, and the engines (5,000 dollars F.O.B. Los Angeles) were delivered with blowers unless otherwise specified. Because supercharging still was a mystery to most of the racing fraternity, the Miller 91 also was designed and sold at first in unblown form, equipped with the familiar four twin-throat Miller carburetors. Early in the formula these began to be replaced by the superior Winified carburetors and the same was true of the single-throat Miller carburetors with which the supercharged 91's were fitted.

It was immediately after the 1925 Indianapolis race that the Miller team went to work on the design of the 2.1875 by 3.091 engine and chassis, as well as crankshafts, cylinder blocks, and other components for the conversion of 122's to the new displacement limit. Fuel still was leaded gasoline, and during 1926 Miller announced the following performance figures, obtained on his Italian Ranzi dynamometer and running a compression ratio of 5.35 to one:

BHP	RPM
118	5,500
136	6,000
150	6,500
154	7,000

The Duray front-drive after restoration. The gear tower rests against the cylinder blocks. Behind them are crankcase, then intercooler and supercharger. IMS

The Miller 91 achieved a high degree of reliability which was remarkable in view of the outputs for which it was originally designed and which it ultimately achieved. The engine, complete with blower, inlet manifold, and carburetor, but without magneto, weighed from 292 to 330 pounds, this spread being due to the progressive beefing-up of many parts and the substitution of bronze for aluminum in blower drives and housings. The tiny aluminum crankcase resembled filigree. It had little strength of its own, and it was debated whether the case could have any significant stiffness at all if it were not for the crankshaft and main-bearing diaphragm assembly which helped to hold it together. Originally it had very light cast-aluminum covers which were bolted over the access holes in its sides. To stiffen the case Miller adopted thick, heavy, deeply ribbed side plates. Even these were not strong enough for many racing men in the field who used quarter-inch iron "boiler plate" instead. And the case itself was steadily beefed up by Miller, who had wood shaved off the core pattern until the thickness of many parts of the crankcase was increased by over 100 percent.

Indicative of the refinement embodied in the Miller 91 was the mere .03-inch clearance between the crankshaft counterweights and the sides of the crankcase; the same clearance was provided between the counterweights and the studs by means of which the main-bearing diaphragms were retained. This precision and economy of space and material was consistent throughout the watch-like engine and had much to do with its high cost.

Countless improvements, large and small and from the factory and from the field kept boosting the output and reliability of the 91. Among many other improvements which he contributed, Lockhart cured the engine's tendency to have its wrist pins split its connecting rods like kindling wood and for its valve stems to fracture. He also introduced the supercharger intercooler. In 1927 Duray, using Miller's dynamometer, began the experiments with alcohol fuel that were to bring about a great leap

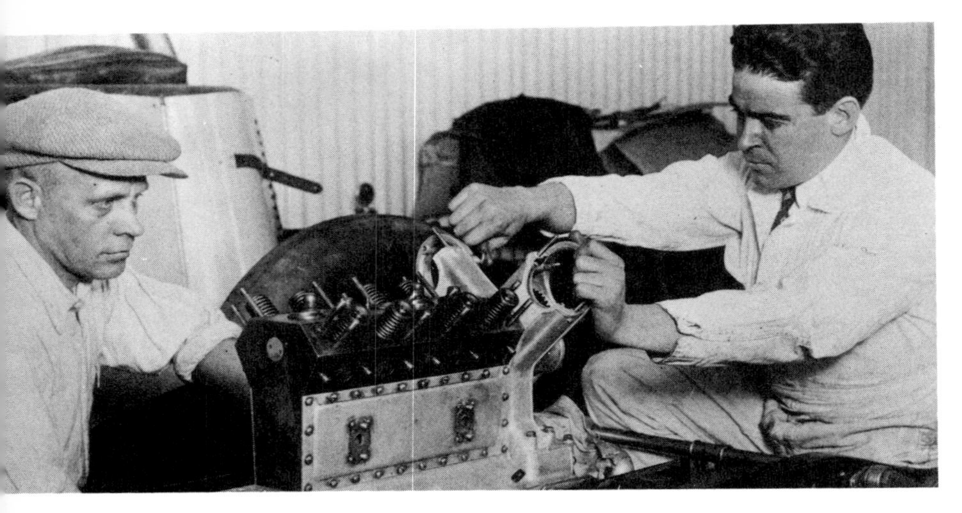

Good drivers usually were excellent mechanics, while the best mechanics were men having consummate skill and the guts to ride in the machines that they prepared. (Left to right) Ernie Olson and Bennett Hill assembling the gear tower to a 91 on which only the front cylinder block has been mounted. TED WILSON

in both reliability and power output. By 1928, and using an intercooler of his own design, Duray was running a 9 to 1 mechanical compression ratio which, with blower boost, calculated out to an effective overall compression ratio of about 15 to one. From this combination he claimed to realize over 265 BHP. If that figure seems high, so do his phenomenal record speeds.

To our knowledge the pinnacle of Miller 91 performance was achieved by Frank Lockhart. After his Muroc Dry Lake 164-MPH record it was simple enough for Zenas Weizel to make an accurate, scientific extrapolation of the urge required to produce the results. It was 285 HP, at 8,100 RPM—a quite fantastic figure.

Only the best mechanics were able to approach Lockhart's combination for ultimate performance, but it soon became commonplace to pull a reliable 230 HP from the Miller 91. It cheerfully turned 8,000 RPM (a piston speed of 4,000 feet per minute) and some of the most highly tuned machines would wind out to 8,500 RPM. Their screaming sound can only be imagined. The spur-gear drives to camshafts and supercharger produced a consistent hackle-raising shriek. The exhaust note depended upon the pipes into which the exhaust manifolding fed. Some favored a single, four-inch pipe, which produced a soft, deep boom. Others preferred a split manifold which fed into a pair of two-inch pipes. They gave off a ripping-canvas sound to end them all. Including marine racing versions, about fifty Miller 91's were built.

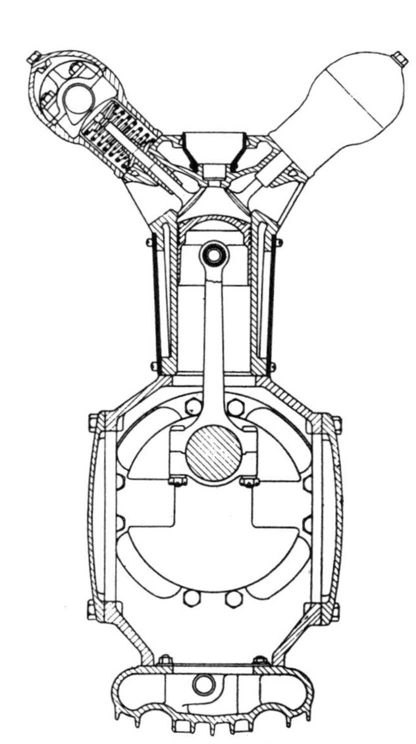

Left: The 91 in transverse cross-section. JOHN R. BOND Right: valve data for the Miller 91, from Goossen's original notebook. LEO GOOSSEN

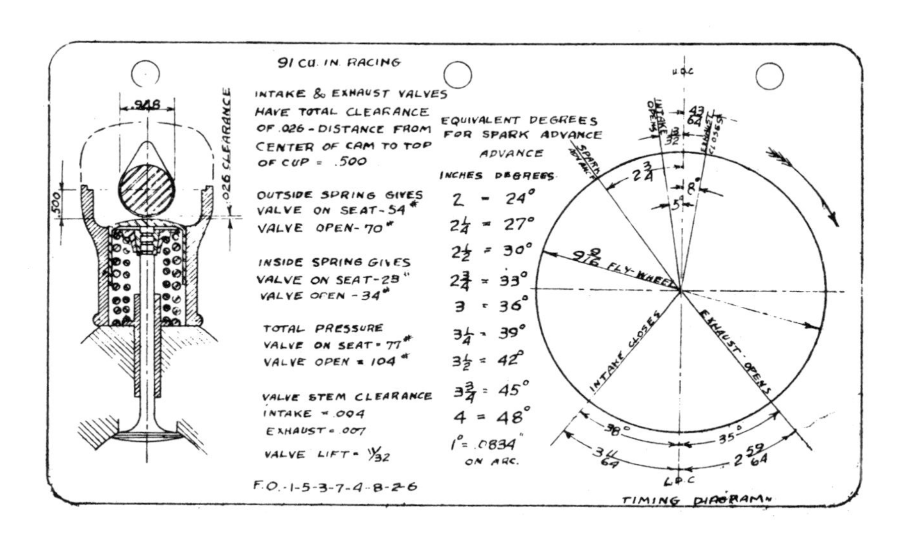

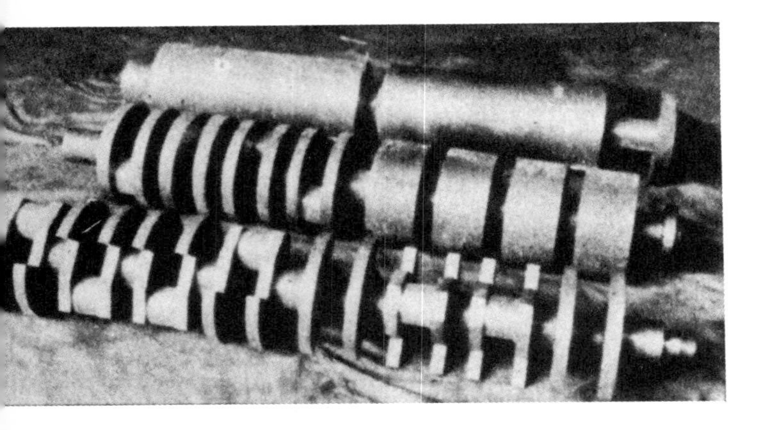

Left: Stages in the machining of a crankshaft from a steel billet. GB COLLECTION
Right: Barrel crankcase for four-cylinder engine. *Car & Driver*

In 1926 famous boat racer Gar Wood commissioned Miller to build him a 310-cubic-inch straight eight for Junior Gold Cup and 340 Class Hydroplane racing. It was basically a scaled-up 122 which developed 200 BHP at 3,500 RPM. Out of this engine Miller developed a 620-cubic-inch V16 for Gold Cup boats and for 725 Class Hydroplanes; its output was 425 BHP at 3,500 RPM. Also from the 310 Miller derived his Marine 151, a 3.406 by 4.125, 150.4-cubic-inch, four cylinder for 151 Class Hydroplanes and speedboats. With an output of 102 BHP at 4,000, or 160 BHP supercharged, this power plant ruled the water as the 91 ruled the boards and bricks. It sold for only 1,100 dollars unblown and 1,500 dollars blown. Enlarged to 183 cubic inches in 1930 for track use it became the foundation of the entire dynasty of Offenhauser fours.

With the exception of very few components, Miller race cars were built in their entirety in the plant at 2652 Long Beach Boulevard in Los Angeles.

The frames were hand-formed from .125-inch flat sheet stock of mild steel and had a channel section. Viewing a 91 front-drive frame rail in transverse section there was, starting at the top, a one-inch vertical lip. The upper horizontal surface was 3.25 inches wide, the main vertical surface was five inches deep, and the lower horizontal surface was 1.5 inches wide. The rear-axle kickups were hammered out by hand over precisely finished cast-iron forms and the finished rails showed no tool marks whatsoever, such was the quality of the workmanship upon which Miller insisted. Most of the rear-drives had three frame cross-members, and the rear engine mounting also served in this capacity. In the front-drive there was only one cross-member as such, this being a tube at the extreme rear of the frame. The front-drive housing, the rear engine mounts, and the forward fuel-tank mounts doubled in this capacity.

The tail sections of some rear-drive Miller cars were aluminum shells which enclosed much smaller fuel tanks, but usually tank and tail were one and the same

Crankshaft and main-bearing webs or diaphragms for four-cylinder engine. GB

and were part of the vehicle's basic architecture. In the case of the front-drives, the faultlessly smooth tank was hand-formed from steel sheet. At the front and near the bottom a 1.125-inch steel tube pierced the tank transversely and was riveted to a pair of large, ribbed, bronze brackets which, in turn, were riveted to the inner surface of the tank. Each end of this steel tube terminated in a ball which bolted into a bronze socket which was bolted inside the frame channel. In spite of their concealed location and typical of Miller workmanship, these socket castings were filed by hand to a smooth finish. When the front tank-mounts were bolted in place, the cross-tube became a fourth cross-member of the frame. Another ball-joint was bolted to the center of the rear cross-member to provide a flexible three-point mounting for the tank. Its internal surfaces were tinned to prevent corrosion and elaborately baffled to prevent surging of the fuel. Miller originated his own large, positive-locking, cam-and-lever fuel-tank cap which still is standard in American professional racing.

On the front-drive the 4.5-gallon reservoir for the dry-sump lubrication system was a steel tank with a finned cast-aluminum base and was carried just under the cowl. It, too, was mounted at three points, being bolted at each of its front corners to the cowl-firewall frame and riding at the rear on a ball-and-socket joint.

The cowl frame consisted of .625-inch square steel tubing welded to light angle-iron rails which rested on the frame rails. This tubing and the radiator were the only supports for the .075-inch aluminum cowl and hood panels. Scarcely any welding was used on the entire vehicle.

Miller's honeycomb copper radiator cores were made by the Eagle Radiator Company of Los Angeles, and through the 122 period used separate radiator shells in the conventional manner. At the outset of the 91 program he adopted the type of radiator in which the top header tank serves an ornamental purpose as well as a functional one and which therefore provided more efficient cooling. In the case of the front-drives, the bottom header tank took the form of an aluminum casting which was bolted to the underside of the front-drive unit. It communicated with the radiator by means of a 1.125-inch brass tube which passed through the front-drive unit and the casting itself often included longitudinal copper tubes for the passage of air. Its purpose was to provide a neat and direct passage for the coolant to the water pump.

Miller was the first to use chrome-molybdenum steel in racing vehicles. He used it first for the front axles of his 122 rear-drives and for the rear axles of his front drives.

The rear axles of Miller's front-drives were startling and controversial because of their 2.5 inch drop-offset at the wheel-spindle ends. This offset—a feature of the 1965 front-drive Oldsmobile *Toronado*—made possible lower frame kickups, lower spring mounting, and lower over-all lines. Miller's offset wheel-spindles were machined from billets of 6145 chrome vanadium steel and then shrunk into their chrome molybdenum axle tubes. Holes were drilled through both offset and tube and then filled with weld, making them an effective unit.

The de Dion tubes of the front-drives were made in three sections: two gracefully curved outer tubes and a removable central one which allegedly permitted easy access to the transmission and final drive. These tubes, including their long spring-perch flanges, were machined from 6145 forgings, then hardened and heat treated. After their elaborate machining the outer tubes were packed with sand and their ends sealed with soldered-on caps. Then they were heated with a pair of torches and bent slowly over a jig, then hardened and heat treated.

Miller's feather-light male wheel hubs (3.5 pounds for the rear-drives) were machined from massive forgings. The rear wheels of the front-drives used standard

Left: Clutch components of the 91. Three-pin positive-locking element is at right. Below right: Total weight of 91's worm-and-wheel steering gear was 15 pounds. Below left: 91 steering linkage components—jewelry. GB

52 millimeter Rudge-Whitworth or Wire Wheel Corporation of America hub shells. The front hubs, because of their universal joints and oversized steering knuckles, required more space. For these Miller designed and machined his own 62 millimeter hubs and wheel-hub shells, to which standard 20-inch rims were laced with wire spokes in the unusual multiple of seven. The lock rings and wheel-rim lips often were drilled for cotter-pinning to prevent loss of rings and tires in the event of a blowout.

Miller made his entire drive train: clutch, three-speed transmission, universal joint, drive shaft, torque tube, final drive, differential, and rear axle housing. The latter consisted of tapered steel tubes bolted to a light-alloy center section. The multiple-disc clutch employed discs made from the finest buzz-saw blades and was equipped with steel dowels which provided a positive mechanical coupling when the clutch pedal was fully released. Ball bearings were used throughout the transmission and, in fact, throughout all Miller cars. The only exceptions to this were the connecting rods and crankshaft bearings (the front main bearing always were merely a bronze bushing) and, in the straight eights, the nine bearings for each camshaft. These bearings were machined in the aluminum of the camshaft housings. They were perfectly reliable and not subject to abnormal wear.

Miller's leaf springs were made of high-carbon spring steel by the United States Spring Company. The leaves were oil-quenched at about 1,750 degrees F., then drawn in a furnace at about 1,100 degrees F. to give them the required resilience. All this produced hard scale on the springs which could only be removed by grinding, the

Left: brake drums were nickel plated, shoes were of cast aluminum, and hubs were machined from solid-steel billets. GB

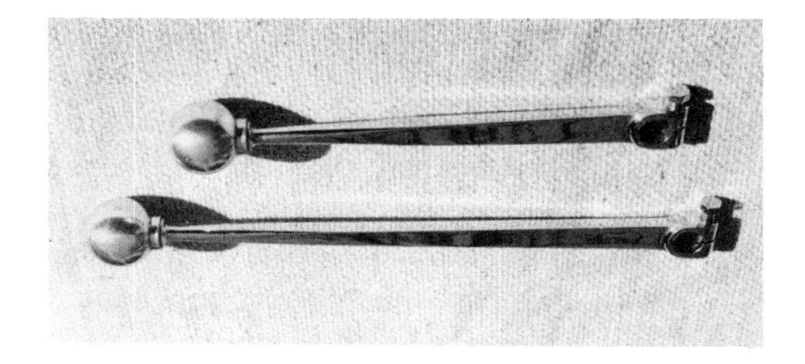

Right: Brake and shift levers were machined from fine bar stock, with rectangular section blending into circular. Aluminum knobs were thin-walled, cored castings. GB

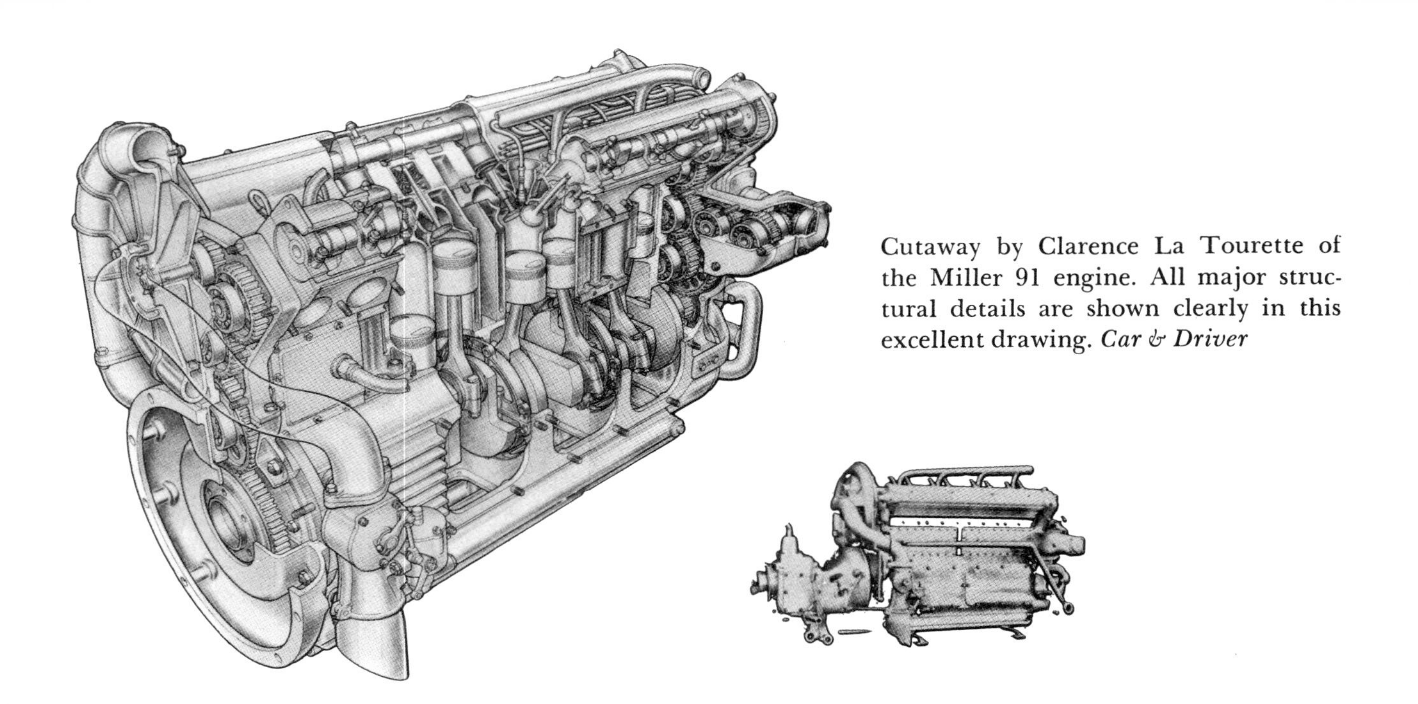

Cutaway by Clarence La Tourette of the Miller 91 engine. All major structural details are shown clearly in this excellent drawing. *Car & Driver*

results of which were the only instances of slightly rough finish on Miller cars. The springs were gun-blued and lubricated with graphite-impregnated grease. This blued finish was used on many parts of some Miller 183's and 122's, including their radiator shells; each wire of each grille was bent over a form and hand-soldered to the shell.

Miller made his own steering machinery. The gears were of the classic worm-and-wheel type, with an important difference. While the conventional form would develop irreparable backlash with wear, Miller's gear wheel was split on its radial center-line and bolted together. Minute adjustments of the gear-wheel halves relative to each other could compensate for the slightest slack in the system. His resilient, four-spoke steering wheels were made on the premises and their wooden rims were fitted by his patternmakers.

Miller's brakes were intended to serve only for occasional pit stops and, as such, were much better than they had to be. They were cable-operated mechanical-servo units with strong bronze backing brackets and light alloy shoes. The foot pedal operated the rear brakes and the hand lever the front brakes. Brake lining normally was of brass-wire impregnated material of great durability and total lining area was lavish: 285 square inches in the case of the ex-Ralph Hepburn front-drive. Still, when Miller made a 13-inch brake drum it weighed only nine pounds and therefore expanded very rapidly under use. They were designed for speedway racing and, for that use, they were excellent.

Miller's masterful touch showed in every piece of his machines. For example, he could have made his hand-brake and shift levers out of ordinary bar stock. Instead, he machined the stock to a rectangular cross-section with rounded edges which tapered and evolved into a circular section as it approached the flange that supported the thread for the knob. And the knob itself was a cored, thin-walled aluminum casting. Even in this trivial component Miller wasted not a gram of weight, while making it a thing of beauty and wonder.

PART FIVE

Pinnacle of the Golden Age

chapter 22

SUPERCHARGING

TO SUPERCHARGE AN ENGINE is to force more fuel-air mixture into it than it can inhale with atmospheric pressure alone. The extent of American achievement in this field is not generally known.

The positive displacement blower, using a pair of hour-glass shaped rotors, was invented by Philander H. and Francis M. Roots of Connersville, Indiana, in 1859. They designed it with a new type of "water wheel" in mind, but it failed in this capacity. The following year they tried it as an air blower in a foundry, met with success, and received their first of many patents in this field on September 25, 1860. Because its pumping efficiency is high over the entire RPM range this type of supercharger was best suited to road racing and dominated the Grand Prix sport in Europe from 1924 until 1951. Discriminatory regulations ended the supercharger's reign.

In France in 1902 Louis Renault took out a patent on a centrifugal fan which blew air into an engine's carburetor intake; nothing came of the idea. But in 1906 in Pottstown, Pennsylvania, Lee Chadwick of the car manufacturing firm which bore his name fitted a legitimate centrifugal supercharger to one of his Great Chadwick Sixes. He was indebted for the idea to driver Willie Haupt, and the eight-inch impeller driven at five times crankcase speed by a three-inch leather belt on the flywheel was a definite success. It encouraged him in 1907 to produce a three-stage blower along the same lines. It blew through the carburetor as the original had done but had a six-to-one drive ratio. Six such Chadwick "production" cars were built. One was timed by the AAA at better than 100 MPH and in 1908 won the local Giant's Despair Hill Climb.

The Duesenberg "sidewinder" supercharger. Pete de Paolo at the wheel of his 1925 mount. JERRY GEBBY

At the opening event on the Indianapolis Speedway in 1909 Len Zengle won the ten-mile free-for-all race on a Chadwick that had just won the Algonquin Hill Climb. His speed was second only to that of Oldfield's Benz, and his was referred to as "the most hair-raising performance of the afternoon." After having built 264 cars Chadwick went out of business in 1911, and his remarkably foresighted idea passed into limbo.

During that same year Marc Birkigt of Hispano-Suiza experimented with a piston-type air pump but abandoned the idea, presumably because it took more from performance than it gave.

Scientific exploration of supercharging began no earlier, it seems, than 1916, in response to the needs of wartime aircraft. Germany led with the rapid perfection of mechanically driven centrifugal blowers and with their application to aircraft of all sizes. Since the plane with the highest flight ceiling has a marked advantage over its enemies, supercharging was a very important factor in the strength of German air power.

Meanwhile, Rateau in France and General Electric in the United States had been working on the same problem. Both favored a similar approach, which was to use the energy of the engine's exhaust gases to drive a turbine-type rotor which, in turn, would drive the rotor of a centrifugal compressor. This was the beginning of the turbo-supercharger. GE's man in charge of the project was turbine expert Dr. Sanford A. Moss.

The turbo-supercharger was in an advanced stage of experimentation when, late in 1917, a Captain Kerr and a Monsieur Le Blanc were assigned by the United States Government to develop a gear-driven centrifugal supercharger for the Liberty engine. McCook Field, where the bulk of this and Moss's work was concentrated, was not far from Indianapolis, and there the Allison Engineering Company, a specialist in high-speed gearing, as well as a manufacturer of Liberty engines, was called upon both to manufacture the experimental units and for technical counsel. These projects were terminated by the Armistice but were reactivated in 1921 by the Engineering Division of the United States Air Service. In other words, there was considerable supercharging know-how in what had become Duesenberg's home town. And it was probably significant that James A. Allison was one of the founding owners of the Speedway and had an avid interest in all aspects of racing.

In late 1919 Mercedes in Germany had begun experimenting with the supercharging of automotive engines. The Roots-type blower was soon chosen as the one best suited to use over the entire RPM range, and a car so equipped and driven by Max Sailer won the 268-mile Coppa Florio in 1921—the first supercharged victory since Chadwick.

As we have seen, the little Duesenberg plant at Elizabeth, New Jersey, became a facility where practically every engine of interest to the United States Government was brought for testing, and it seems to have been during this period that Fred and Augie first met General Electric's Dr. Moss and were first exposed to centrifugal supercharging. They immediately saw beyond the conservative aeronautical approach, which was merely to maintain the pressure of one atmosphere on the intake system up to a given high altitude. They thought in terms of pouring all the pressure to the engine that its structure would stand. Moss of course was intrigued by the experimental possibilities and offered his help.

Duesenberg began probing this unknown field with a small Moss blower with a five-inch impeller. Drive by means of exhaust gas was out of the question for the results he sought, and he had to design and develop his own mechanical drive. The results were slow in coming, as Eddie Miller recalled:

> Dr. Moss came by a few times over a period of a year or so, just to see how we were doing. He didn't seem to know much more about it than we did. It was just a matter of getting bolder and bolder and running them faster and building them larger.
>
> So you enter this new world of the supercharger. You wind it up and nothing happens. You wind more and more and finally you get two or three pounds of pressure and you think that's terrific. Now, if you were just getting by on the compression ratio you had before, it would start knocking and detonating and you'd burn a hole through a piston in nothing flat. It took two or three races when you're having this kind of trouble to figure out what's going on. Finally you do and you lower the compression ratio and boost the pressure, maybe to 4.25-to-one and ten or twelve PSI. You'd almost double the size of the charge by kicking the pressure up to almost an atmosphere, and there was the power at last.

Duesenberg had superchargers in action by 1920. The Richards Special, which John Boling drove to eleventh place at Indianapolis that year, was the noisiest car on the track, but only Fred's "family" knew why. The device was kept a secret while its perfection was pursued. It still was full of problems when at Indianapolis in 1924 Joe Boyer leaped into the lead on the first lap, "like a flash of light, a rush of wind." The blown Duesey won, if only by seconds, the Technical Committee verified its engine's displacement, and the secret was out.

That was on May 30, and on June 8 Duray thoroughly whipped de Palma at Ascot with his sparkling new "1925 model supercharged Miller," shattering the lap record in the process. Another major race was to be held on June 22, but all entrants went on strike, refusing to run unless Duray removed the device, which he finally agreed to do. He also won this race, hands down. How had Duray and Miller mastered in days what had taken Duesenberg years?

I recently asked Dick Doyle, Duray's mechanic at the time. There *was* no supercharger; that was said just to psych the competition and electrify the press. As for the car being the next year's model, it actually was just Ira Vail's historic, original, wrecked Miller 183, which Duray and Doyle had rebuilt with no money but enough skill to make fools of the experts.

The day after the Indianapolis race Miller and Goossen had gone directly to Allison Engineering for whatever information they could get on centrifugal supercharging. Other aircraft firms were consulted, and the technical literature was searched. After that it was cut and try, but Miller lost little time in becoming a past master of supercharging technique.

In basic principle the centrifugal supercharger is a simple device, similar to the water pump in most cars. The major moving part is a finned disc, the rotor or *impeller*. It is contained in a housing which surrounds it with a narrow *diffuser* passage which opens out into a wide, spiral *collector* space called the *volute* or, colloquially, the "snail." The volute's cross-section increases gradually toward the

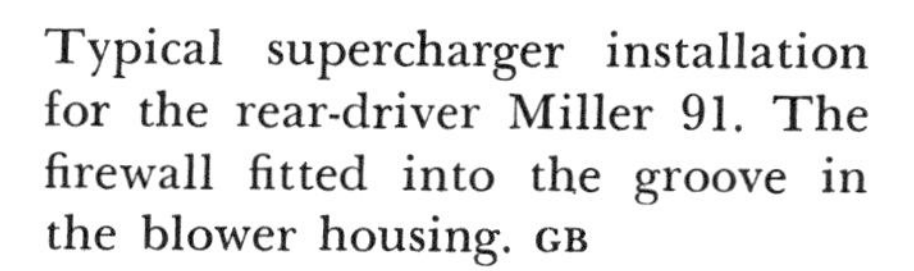

Typical supercharger installation for the rear-driver Miller 91. The firewall fitted into the groove in the blower housing. GB

Left: Intercooler of the 1928 Miller-Boyle Valve Special followed the basic Lockhart pattern. GB

Below: Inducer vanes first were carved as integral parts of the impeller, then machined separately. GB

Below: Zenas Weisel examines the Duray intercooler in the author's workshop. GB

Below: Supercharging was a black and unknown art that produced an immense variety of impellers. GB

discharge end. The spinning impeller draws air in at its center or *eye* and flings it into the diffuser space. Here the air's velocity drops and its kinetic energy is converted into pressure. The volute serves to collect the compressed air and lead it to the discharge opening with a minimum of loss. The only rubbing parts involved are bearings and, if required, drive gears.

Since velocity is converted into pressure by this type of pump, the velocity of the impeller tips is critical. Unlike the Roots type, which delivers useful boost from very low RPMS, the pattern of classic American racing blowers was to give nil boost at 2,500 RPM, about nine PSI at 3,000, and twice that at 6,000. This low-end inertness of the centrifugal blower completed the schism which had been developing between road-oriented European race cars and oval-track-oriented American ones. But still the Americans made a contribution to the art of supercharging that all the world could use. This was Duesenberg's innovation of locating the carburetor on the intake side of the blower rather than on the discharge side, which was the then-general fashion. Thus the fuel's latent heat of evaporation could reduce the heat of the incoming charge, and the carburetion system was simplified.

Designing a centrifugal pump of tolerable efficiency was no great problem, but driving it presented challenges, the most severe of which was driving the impeller reliably. With peripheral speeds of the order of 1,300 feet per second, 78,000 feet per minute, 890 MPH, and shaft speeds of 40,000 and over, sudden changes in engine speed tended to tear the system apart. Deceleration of the reciprocating engine parts was the most damaging because it was more abrupt than acceleration.

Duesenberg had found a happy solution after long and costly trial and error. His straight eights had been laid out originally with a power takeoff on the right-hand side of the engine; it ran horizontally from the front timing-gear case back to drive the distributor. It was a simple matter to put a skew gear on this shaft and drive a blower driveshaft which ran transversely in the space that existed between number four and number five cylinders. The amidships blower, mounted alongside the crankcase on the left-hand side of the engine, was cushioned against sudden changes in engine speed by this transverse shaft which also acted as a torsion-bar spring. This side location of the supercharger caused blown Duesies to be known as "sidewinders." To drive his impellers Fred used a three-pinion planetary gear arrangement that was closely similar to that which was used at the top of the Model T Ford steering post. It had the great virtue of permitting gear ratios to be changed very simply and many experts in the field considered it the best blower drive of all.

Miller had no such ready-made basis for a blower drive on his 122's although he thought he had in harnessing the camshafts themselves. His first blower was mounted transversely, high at the rear of the engine. The impeller pinion was driven by a pair of spur gears which, in turn, were driven by tongues inserted in the ends of the camshafts, but the tongues broke constantly. Miller corrected this by equipping the camshafts with gears which meshed with the two drive gears. This approach proved durable, but it was claimed that valve timing was spoiled when impeller inertia made the camshafts act as torsion bars.

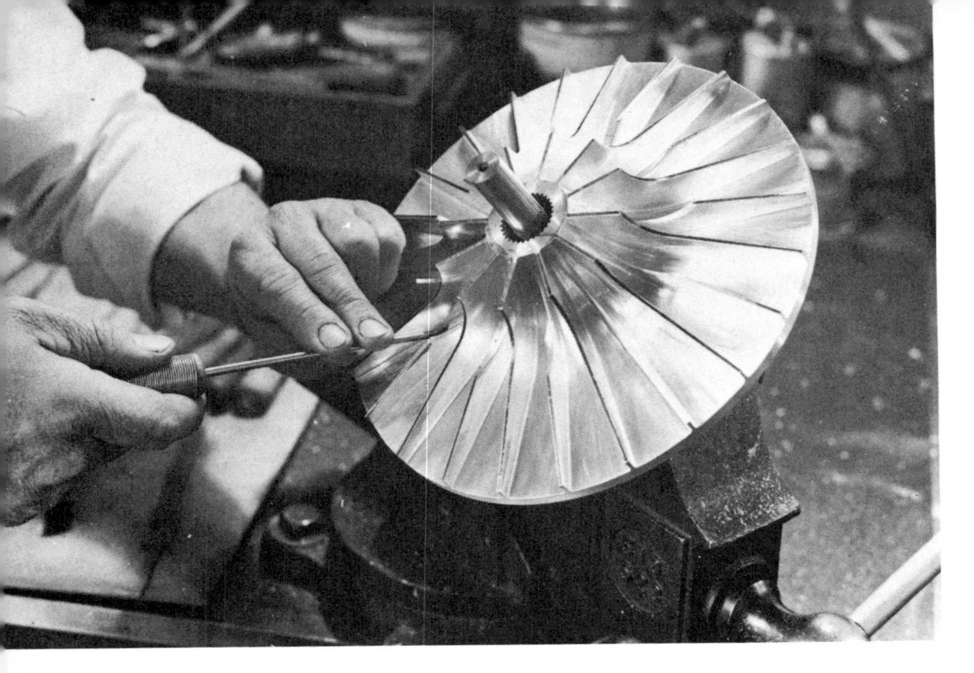

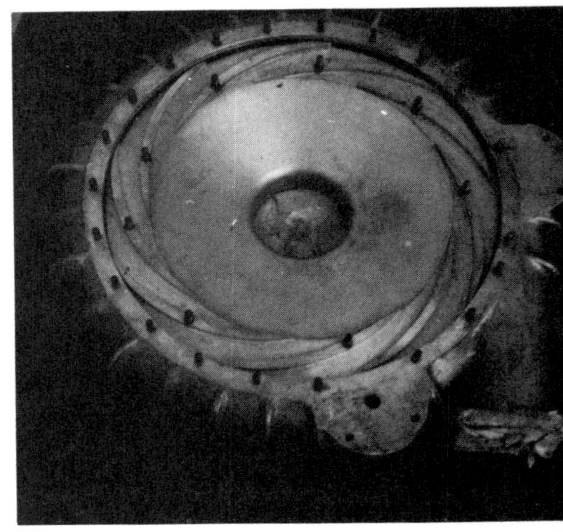

Among other things, creating a good impeller is a sculptor's art. Right: Inner face of volute or "snail" of same blower. Note diffuser vanes. GB

These problems were resolved with the first Miller 91's. In them the blower was mounted on top of the bell housing at the rear of the engine and was driven by means of a spur gear which was bolted between the flywheel and the crankshaft. This gear had a two-piece hub which was made flexible first by means of flat springs, then by rubber blocks and, finally, by coil springs. These worked best and the cushioning problem was resolved for Miller.

Miller's impellers were machined and hand-sculptured from duraluminum billets, occasionally from billets of magnesium and even from vanadium steel. The first Miller impeller had a diameter of 5.5 inches, but this dimension gradually was increased to seven and finally to nine inches. His first impeller had six vanes; he experimented with as many as eighteen but found the best results with twelve.

He quickly adopted bladed diffusers which, by mechanically directing the air in the optimum direction, permitted the diameter of the blower casing to be reduced to the minimum. He also was quick to add inducer vanes to the eye-area of the impeller. These are extensions of the main vanes and are curved to meet the incoming air at an angle that eliminates the formation of undesirable eddies; they scoop the air rather than slap it. First they were made integrally with the impeller and then, for economy and simplicity, were fabricated as a separate, small impeller which bolted onto the main impeller.

The original Miller blower and gear-drive housings were aluminum castings which weighed almost nothing. As rotational speeds and blower pressures moved toward their occasional peaks of 45,000 RPM and 30 PSI, failure of these housings became increasingly common until both were replaced by extremely heavy bronze castings. Among other gains, these put an end to blower housings popping like bubbles in response to backfires that no popoff valves were quick enough to relieve completely.

Said the World War II Report L-341 of the National Advisory Committee for Aeronautics:

Miller-91-based Detroit Special was financed by Cliff Durant and designed by Milton and van Ranst. Dr. Sanford Moss was responsible for the pioneer two-stage blower. IMS

A supercharger . . . in compressing the air, also raises its temperature. The temperature of the air must then be reduced in order to increase its density and to prevent the detonation that results from high air temperature. The reduction in temperature is accomplished by an intercooler, which is simply a heat exchanger in which both fluids are air.

The genesis of the intercooler in American automobile racing is described in Chapter 24. After Frank Lockhart exploited the principle in victorious secrecy for nearly a year it was adopted by Duesenberg and by many private owners of Miller cars, who designed and built their own intercoolers, thus relieving Miller of the need to emulate his former protégé.

The howl of the supercharger was the characteristic cry of the American thoroughbred car at the peak of its development. It inspired a bevy of centrifugally-blown passenger cars, among them Auburn, Cord, Duesenberg, Franklin, and Graham. All these used impellers that gave only a slight but useful boost. It was boost up to two atmospheres that put the final brilliance into the Miller and Duesenberg 91's, and it was the pioneer intercooler that permitted such tremendous pressures to be used.

Duesenberg eventually tried this scarcely inspired type of intercooler. He was still using screw-on radiator caps in 1929. Note rear spring mount covering the rubber blocks. IMS

chapter 23

FRONT-WHEEL DRIVE

FRONT-WHEEL DRIVE IS AS OLD AS the self-propelled vehicle and was a feature of Cugnot's historic steam carriage (only one front wheel) which lumbered over the boulevards of Paris around 1770. Overseas automotive designers Humber, Latil, de Riancey, and others tried and abandoned front-drive prior to 1900 due, chiefly, to the universal-joint problems associated with wheels that must be sprung, steered, and driven.

In New York in 1904 J. Walter Christie, who had prospered as a designer and builder of military gun turrets, took out patents on a highly original method of driving an automobile through its front wheels. The chief feature of the design was its mounting of the engine transversely at the front of the vehicle. That same year Christie built a running prototype and in 1905 founded the Christie Direct Action Motor Car Company for the manufacture of taxis which, in general layout, bore an almost uncanny resemblance to the modern Issigonis-designed BMC Mini.

Between 1904 and 1909 Christie built seven front-drive race cars which were powered by twins, fours, V4's, and V8's of his own creation. They were notoriously brutal to handle but won their share of speed contests for years after Christie had given up any commercial hopes for the idea. In 1907 he set a new record for the dirt-track mile at 69.2 MPH. On Dec. 18, 1909, on the newly brick-surfaced Indianapolis Speedway he set a record for the flying quarter-mile at 97.6 MPH. Then Barney Oldfield acquired the biggest 140 BHP Christie and barnstormed it from coast to coast

1910 Christie front-drive with huge V4 engine laid almost flat. Note sophisticated front suspension and integrally cast wheel hubs and spokes. IMS

for years. In 1915 on the Tacoma boards he averaged 113.9 MPH for a mile, and at Indianapolis in 1916 he set a new lap record of 102.6 MPH. Ed Winfield saw Barney raise the AAA dirt-track mile record to 77.6 MPH at Bakersfield, California, in 1913. The sight, he said, was terrifying. The machine would barely steer, and it bolted around the track in a series of rock-and-dirt spraying leaps which coincided with each power impulse. Christie returned to the military ordnance field, and immediately after World War I devised a radically new track-drive and suspension for tanks. The system did not impress the United States Army, and Christie was forced to seek overseas markets. His track suspension made possible the modern high-speed tank which revolutionized warfare in the 1930's and 1940's. France, Germany, Great Britain, and the U.S.S.R. all adopted it.

When Jimmy Murphy handed Miller an order for a front-drive race car early in 1923 it was only natural that the Miller team would start thinking in terms of the record-breaking Christie machines. Goossen sketched a chassis with a straight-eight engine mounted transversely in front of the front axle. He was pondering how to get the power to the wheels when Murphy and Brett looked at the drawing board and refused to accept that approach.

The next link in this evolutionary chain is rather mysterious. Between 1918 and 1922 a small designer-manufacturer, Ben F. Gregory of Kansas City, Missouri, built a series of about ten automobiles, all utilizing front-wheel drive. Some were touring cars, some race cars, but all used a de Dion front-axle arrangement. They also used conventional engines mounted longitudinally in the frame, but with the flywheel end and the transmission at the front. For a couple of years Gregory barnstormed dirt tracks and county fairs in the surrounding area as a stunt driver. His mount was one of his own front-drive single-seaters, powered by an OX5 and later by a Hispano-Suiza aero engine. In 1921 he exhibited a front-drive tourer at the Kansas City Auto Show.

Gregory reported having had several contacts with Miller in the early 1920's. He certainly anticipated much of the basic arrangement which Miller was to carry to a remarkable degree of perfection. He deserves recognition which he has never received.

The greatest virtue of transverse engine mounting in a front-drive vehicle is the concentration of its weight upon the driving wheels, in the interest of optimum traction. After Murphy's rejection of this layout Miller and Goossen sketched a Gregory-type layout. Then they laid out several conventional transmissions of two and and three forward speeds, but these were so space-consuming that they forced the engine far to the rear and thus reduced all-important front-end loading.

They solved this problem by designing a transverse transmission which nested between the frame rails and as closely as possible to the flywheel; the space between the rails was just seventeen inches.

A multiple-disc clutch with four friction faces was assembled within the recessed flywheel. Because Miller had had trouble with clutch slippage on his early 122's he had developed a positive pin drive, as mentioned in Chapter 21. When the clutch pedal was released gradually the friction surfaces would grab slowly until the three steel dowel pins dropped into their corresponding holes, locking the whole system mechanically. This clutch became standard for both front and rear-drive Millers and

The same car being campaigned by Barney Oldfield, in foreground. GB COLLECTION

required skilled manipulation. To apply power before the pins had bottomed was to destroy the mechanism.

A driveshaft less than a foot long fit into a splined hub on the clutch pressure plate and projected into the front-drive gearbox. The pinion meshed with a ring gear which, through a rather amazing series of three shafts which telescoped inside each other, fed torque through one reverse and three forward speeds and then through a tiny, jewel-like differential to output joints on each side of the incredibly compact gearbox.

Miller had a lifelong contempt for patents but this *tour de force* of design was an exception. Without Goossen's knowledge he filed on January 30, 1925, and on November 15, 1927, was granted, a patent for this power transmission device. In the patent application its labyrinthine structure was described in labyrinthine prose:

> . . . an improved driving mechanism for a vehicle, which shall essentially consist of a driving shaft, a driven shaft, an axle, a first sleeve disposed about the axle and rotatable thereabout, a second sleeve rotatable about the first sleeve, a ring gear mounted upon the second sleeve and engaging the pinion so as to rotate the second sleeve, a counter-shaft driven by the second sleeve providing a speed change transmission operatively connecting the second sleeve to the first sleeve, and a differential which shall connect the first sleeve to the sections of the axle.
>
> Because of the slow rotation of the counter shaft, it is possible to shift gears . . . without throwing out the clutch without a very great danger of stripping the gears.

This watchlike mechanism was a brilliant solution to the problem, but it was fragile for two reasons. One was that the transmission gears were on the output, high-torque side of the ring and pinion instead of between the engine and final drive. The other was that the transmission gears—which should have been greatly oversized to handle the multiplied torque—had to be extremely narrow in order to fit into the restricted space available.

This transmission had no right to perform as well as it did. It was reliable in sensitive hands but would tolerate only the most cautious gear changes, and those only for getting up to speed. Most drivers took no chances and used push starts, with the mechanism in top gear. Downshifting was entirely out of the question, but there was no need for it in speedway racing. The conception and execution of this radical, wheels-within-wheels gearbox was only half the job. The other half was getting the power to the bricks and boards.

The backing plates for the inboard front brakes (outboard on the original Murphy car) were machined from heavy bronze and also served as carriers for the sliding-block or square-and-trunnion-pin inboard universal joints (the first couple of front-drives used Thermoid fabric universals). These sliding-block joints were on the inboard ends of tubes which fitted around the axle half-shafts by means of splines and thus formed a sliding joint. The half-shafts, in turn, were assembled to Hooke-type

universal joints. The outer members of these joints ended in splined stub shafts on which the wheel hubs were mounted. The hub and outer universal joint assembly was carried in an exquisitely engineered steering knuckle or spindle yoke which pivoted on king pins which were integral with the fork or yoke of the de Dion tube.

Caster was provided through a four-degree angle which was designed into the spindle yoke. The center lines of king pin, Hooke-joint, and wheel all intersected at the center of the wheel, on the tire center line. Thus, steering had minimal effect on the delivery of power and *vice versa*. As experience with front-drive phenomena grew, Rzeppa and Weiss constant-velocity universal joints replaced the Hooke joints.

The transmission's three speeds and reverse were selected by moving two sliding gears by a remote lever, two long shafts, and a position-selecting device which was strikingly similar to that used on the Corvette four-speed gearbox of the late 1950's.

The central, straight section of the de Dion tube was detatchable to facilitate access to the transmission but all this accomplished was easy removal of the cover of the front-drive unit. To do any important work on its contents required dismantling of the entire, complex front-end assembly.

The components of the de Dion tube of the original Miller front-drive were steel castings, but they proved to be brittle. The tubes of the dozen or so 91 front drives which were subsequently built were machined from solid chrome vanadium hand-forgings.

Commercially available 42-millimeter splined wheel hubs were used at the rear of the front drives, as they were all around on the rear-drive Millers. But the size of the outboard universal joints and their adjacent structures on the front-drives demanded oversized hubs, which were machined from forged billets.

With de Dion front suspension—its first use anywhere on a true race car—inboard brakes, and frame-mounted final drive, the front-drive Millers had very low unsprung weight. Their handling characteristics were very different from those of conventional

Left: One of Ben F. Gregory's front-drives, which used de Dion front axle. Missouri license plate is dated 1920. BEN F. GREGORY Center: The first Miller front-drive. Only this car was built with outboard brakes. GB COLLECTION Right: Sponsor Cliff Durant at the wheel of the Miller Junior Eight. TED WILSON

cars, and some drivers preferred them while others did not. In the hands of those who felt at home with them they were spectacularly superior to rear-drives.

Commenting on the 1926 Indianapolis race, expert Lee Oldfield summed this up nicely at a meeting of the Society of Automotive Engineers:

> I think there is no disputing the statement that the [front drive] cars follow the front wheels and have less tendency to skid, if we assume that the driver has the courage to keep his foot on the throttle. But I can assure you that if the driver takes his foot off the throttle he will immediately have something to think about. Just prior to the "500" Mr. Duray demonstrated that, with the foot on the throttle, the car will travel in the direction in which it is pointed. Pete de Paolo demonstrated that, with the foot off the throttle, the car will not travel in the direction in which it is pointed.

At the same meeting it was pointed out that the front-drives lapped the Speedway with almost constant throttle openings, holding about 6,200 RPM on the straights and dropping to only about 6,000 RPM for the turns. The rear-drives, on the other hand, took the straights at about 7,200, had to shut off early for the turns, and came out of them turning only about 5,000 RPM. Front-drive was heralded as a great engineering breakthrough which was full of promise for the passenger-car industry.

One of the 1926 cars was to play a key role in the evaluation of front-drive. It was the Detroit Special, financed by Cliff Durant and built in the Hyatt Bearing shops in the basement of the General Motors building in Detroit. Its designers were Tommy Milton and C. W. van Ranst. The car was basically a front-drive Miller 91 which they had redesigned to place the transmission on the low-torque side of the final-drive gears. Leon Duray drove this car briefly in the 1926 "500" and found its easy-to-use gearbox a revelation. The car also featured a two-stage centrifugal supercharger which Milton and Van had developed in collaboration with Dr. Sanford A. Moss.

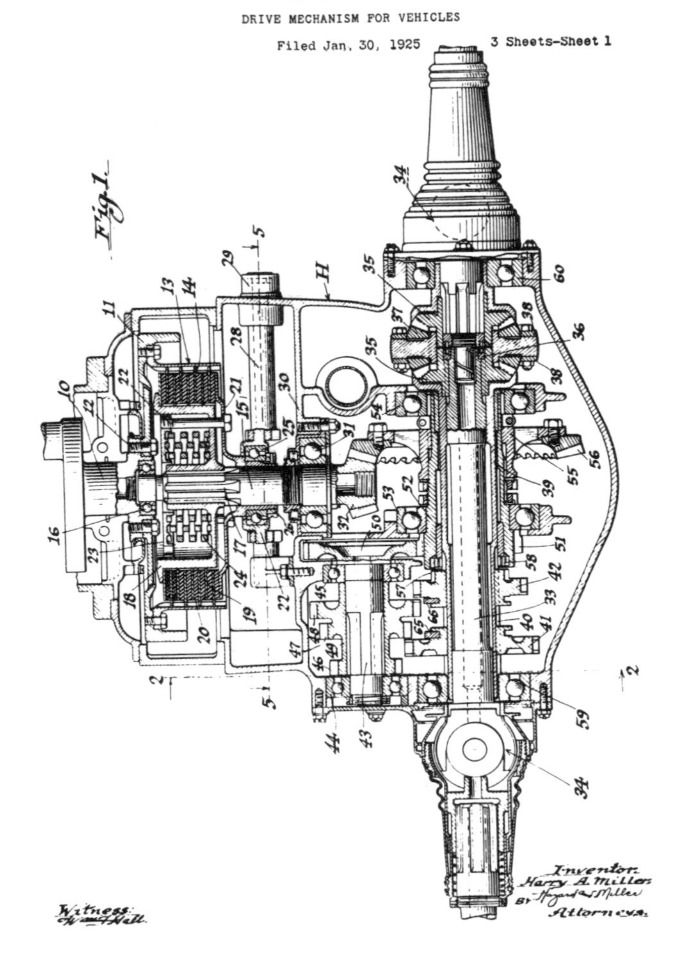

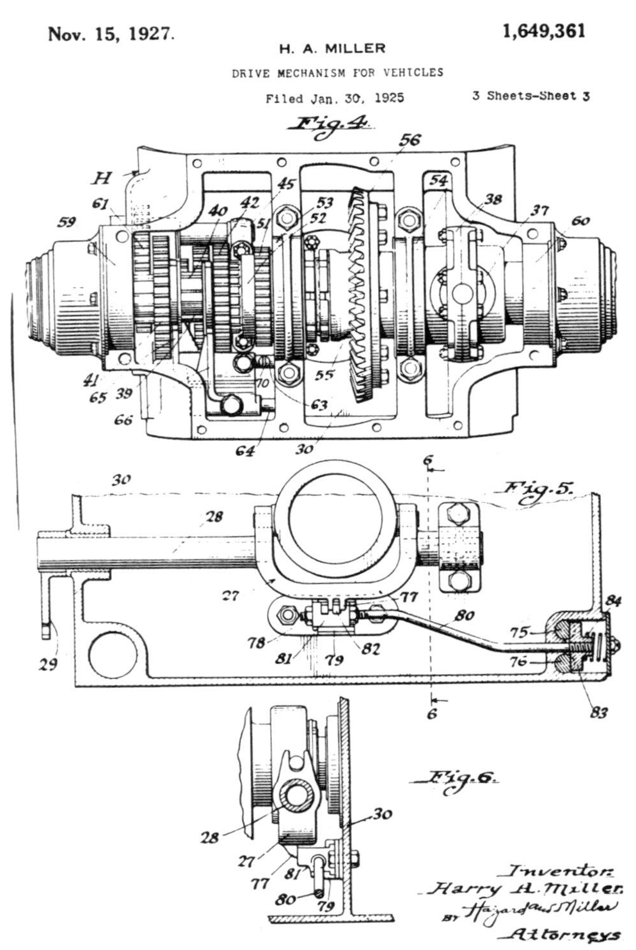

Patent drawings for the Miller front-drive. Inboard brakes were an afterthought. RICHARD FABRIS

E. L. Cord already had contracted with Miller for the design of the front end of the projected L29 Cord car and for the construction of a prototype power train. Duray, who had close connections with Cord, called his patron, told him of the smooth-working gearbox of the Detroit Special, and urged that he hire van Ranst without delay.

Duray's counsel was acted on with lightning speed and, with the title of chief engineer, Van was dispatched to the Miller plant. He recalled:

> When I came out to California on that project I had to use the straight-eight Lycoming engine and we had to use the standard transmission parts. I just couldn't design out of those junk parts anything that I would consider building.

It was Goossen who took hold of the problem and said, "What difference does it make? We've got to build the thing." And Leo laid out most of that front end—for me, actually.

The L29 Cord was America's first front-drive road car to be produced in significant volume. Except for its use of a longitudinal transmission placed between engine and final-drive unit, the L29 layout was closely similar to that of the front-drive Miller race cars, including the de Dion axle, inboard brakes, and quarter-elliptic springs.

Duray's invasion of Europe with two front-drive Miller 91's in 1929 was tied in with the arrival of the first Cord cars to reach Europe. Duray and his cavalcade were met and assisted by W. F. Bradley and by J. A. Gregoire who, in 1926, had begun building and racing, with good success, the front-drive Tracta car. Gregoire drove the new Cords and pronounced them as magnificent in their lines as they were miserable in their mechanism. He regarded their steering as particularly bad and in later years felt that this unfortunate introduction of front-drive in the United States was responsible for its failure to achieve any strong market acceptance there.

Although no front-drive cars had been built commercially in Europe since the turn of the century, Gregoire would not admit to having been influenced by Miller's prior revival of the principle in 1924,* although it was important news in the European automotive journals of the day. There were points of similarity in the Miller and Gregoire designs as well as great basic differences. Certainly, the fullest credit goes to Gregoire for his sophisticated, typically French approach to road-holding and for his own solution to the constant velocity universal-joint problem. DKW in 1928,

* The first front-drive Miller was designed and built in 1923 and 1924. The first photos and descriptions of the car appeared in American automotive periodicals in late 1924. In May and June of 1925, before and after the "500," many photos were published, showing the major structural details of the front-drive system, including views of the front-drive gearbox with its internals exposed. It was, internationally, *the* technical innovation of the year.

Left: Elements of the FD de Dion front axle tube. Center: The dropped rear axle. Right: Another compact touch: radiator recessed to receive carburetor.
GB COLLECTION

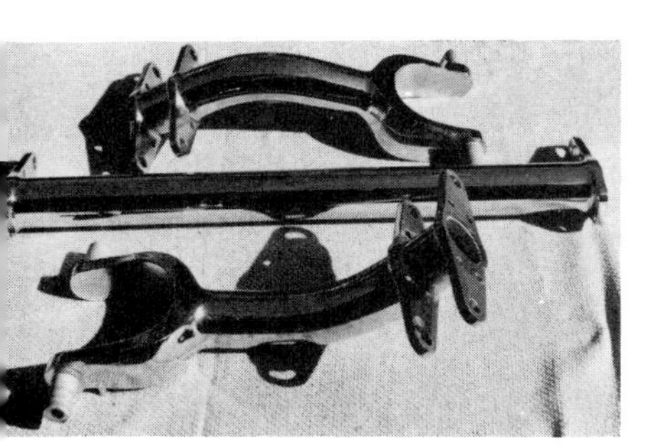

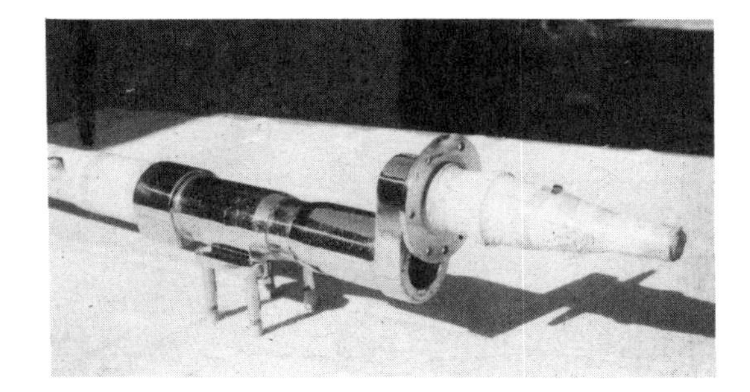

Left: Frame of the ex-Hepburn front-drive Miller 91, with crankcase and engine mounts installed. Center: Same, front view. Right: Clutch installed. Note aluminum spacers between bell-housing and frame. GB COLLECTION

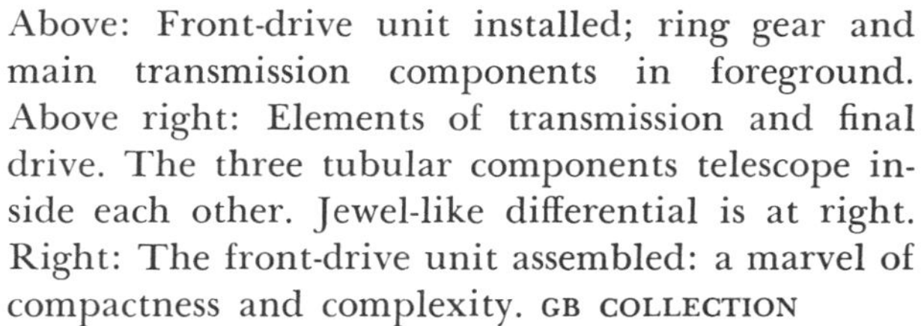

Above: Front-drive unit installed; ring gear and main transmission components in foreground. Above right: Elements of transmission and final drive. The three tubular components telescope inside each other. Jewel-like differential is at right. Right: The front-drive unit assembled: a marvel of compactness and complexity. GB COLLECTION

Left: Suspension components for one wheel. Center: Brakes assembled. Note that radiator bolts solidly to front-drive unit. Right: Brake drums, inner universal joints, male-splined shafts and shock absorbers installed. GB COLLECTION

Above left: Drive line components from inner U-joint to hub cap. Note intricate steering knuckle at left. Left: These components assembled. Above: The whole Miller FD cosmos in harmonious order. GB COLLECTION

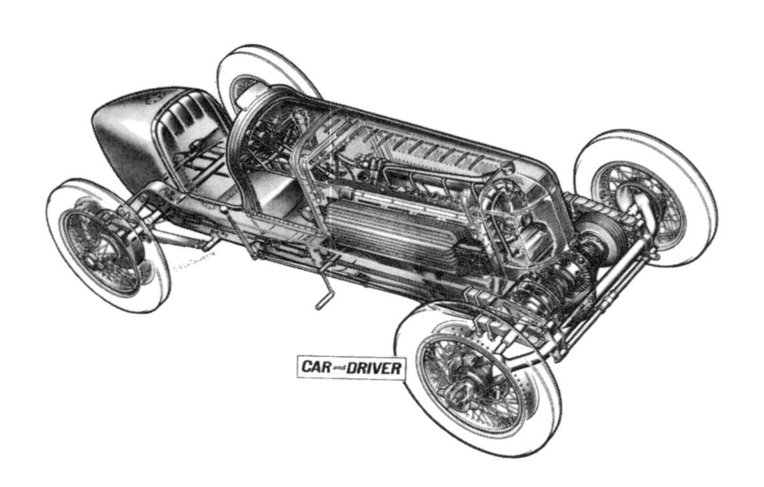

La Tourette cutaway of the Ralph Hepburn-Bugatti-Borgeson Packard Cable Special. *Car & Driver*

Adler in 1932, and Citroën in 1934 were among the manufacturers who brought out front-drive passenger cars, using Gregoire's patents. Miller's United States Patent Number 1,649,361 only covered his marvelous little transmission which, for reasons already stated, no one wanted anyway. But there can be little argument that he, at Murphy's and Brett's instigation, fathered modern front-drive and won worldwide acceptance for it.

Fronty-Ford-powered Hamlin Special had close technical connections with the Ruxton FD passenger car, and was a successful dirt-track campaigner. GB COLLECTION

chapter 24

THE LOCKHART SAGA-I

BACK IN THE ROARING TWENTIES Frank Lockhart was a great national hero. He was one of the greatest racing drivers that the world had ever seen. As a kid just out of his teens and jockeying a homemade bucket of bolts he became famous overnight as "King of the Dirt Tracks." The first time he went near Indianapolis he won the 500-mile race spectacularly and decisively in a car that he had never seen before. When he hit the Championship Trail after that he was consistently fantastic, winning races and smashing records from coast to coast. His name was as much a household word as that of Babe Ruth or Jack Dempsey even before he decided to become the fastest man on earth.

Then he built the Stutz Black Hawk, an American challenger for the world's land speed record. Compared with the British record-holder, Segrave's colossal Sunbeam, the Black Hawk was a mere toy. But experts called it "the most mechanically perfect automobile ever built in the world" and predicted that what it lacked in size it would more than make up for in efficiency.

The eyes of the world were on Lockhart when, at the age of twenty-five, he did go faster than man ever had gone on land. The record for ultimate speed on wheels stood at 206.95 MPH when the wire services credited Lockhart with a phenomenal 225. Then they told how a tire had blown at that speed and how the tiny car rolled and banged eerily down the beach and how a cruel, ironic fate caused his broken body to be flung at the feet of his waiting wife.

Left: Frank Lockhart, "Boy Wonder of the Dirt Tracks." TED WILSON Above: Frank's first and only car prior to his first Miller. MRS. CARRIE LOCKHART

Truculent face and typical sentimentality: "To Mother—Frank S. Lockhart." He carved the radiator shell out of wood. It fell apart. MRS. CARRIE LOCKHART

Where the Lockhart legend survives it has the heroic quality of ancient myth. Illiterate, the hero's intellect was awesome. Obscure, he rocketed to world fame. Penniless, he raced his way to wealth. Frail, his endurance was unmatched. What other men accomplished with wealth Lockhart surpassed with iron will and cold sweat. He surpassed Horatio Alger's most sanguine stereotype, this little guy who slew giants by virtue of nothing but guts, infinite labor, and, above all, brains.

Legend though it is, none of it is far from the truth. Lockhart did all these things. But he did not do them all single-handedly, as the myth relates. Nor were the sacrifices and heartbreaks all his own.

Frank Lockhart stood five foot eight, weighed about 135 pounds, had hazel eyes, and flaxen hair. He was born in Dayton, Ohio, in 1903. It may have been more than significant that the gifted father of Wilbur and Orville Wright lived next door. He liked Frank, let him watch the goings-on in his workshop, and when Frank was three, built Frank a bicycle. The boy rode it well and wrecked the interior of the family home with it. But he spent most of his waking hours in a nearby livery stable watching work being done on the automobiles of the day. At home, he appropriated anything and everything mechanical for his own, from his mother's egg-beater to her sewing machine. He dismantled everything, then tried to reassemble it.

Frank's father, Casper Lockhart, was a theology student from Indiana. He didn't have the mechanical talent to open a pair of pliers, and his ambition was to become a preacher. He died when Frank was six, and his widow Carrie moved with Frank and his brother Bob to California. She took in sewing to keep the family alive. "I had quite a struggle raising those children," she said. "And Frank's demands were very great."

Frank was incapable of playing with other children, had no patience with their aimless frivolity. He just worked on things, built things, and devoured all the reading matter he could find that dealt with things mechanical. He never had time to be bothered with spelling; as an adult he spelled sugar *shougar,* with a certain phonetic logic. Even as a child he loved mathematics and felt far more at home with numbers than with words.

When Frank was eight he built a coaster—a Soapbox Derby sort of thing that all the kids built and raced at the time. He didn't have money for canvas or sheet metal to cover it with, so he used tarpaper. In neighborhood races he almost always won, mainly because he had the power to get the greatest number of kids to push him. Anyone who wanted to associate with Frank had to submit to his will. When his shoes had to be polished Frank took care of one shoe and his brother had to do the other one. When Frank was running race cars no one could drop in at the garage just to watch or talk. He was handed a rag or a tool and put to work. Frank was a tyrant.

He had nothing but trouble in school. He had a mind that was totally his own and would work only at subjects which he considered to be important. They were few. When automobiles still were being built with acetylene lamps Frank frittered away his classroom time drawing cars with envelope bodies. His mother said:

> His teachers were always calling me to school and confronting me with drawings of streamlined cars. They would say, "This is the way your son is spending his time, drawing these ridiculous things." I would protest that those were automobiles. They would say, "What! Those things? Nobody will ever ride in things like that. They're impossible. They don't even have running boards. The child is crazy. Punish him!"

When Frank managed to get himself graduated from high school the principal told Carrie that he "never had been taken advantage of by a student" as he had by her son. The problem was that Frank treated the man as an equal and repeatedly crushed him with superior logic.

By the time he was sixteen Frank had to have a car, family poverty notwithstanding. His first angel was a Jewish vegetable peddler named Gentle. He had a Model T Ford chassis rusting in his yard over in Boyle Heights and offered it to Frank if he would haul it away. At that time Frank felt that anyone who would build a car frame out of steel rather than light hardwood was a fool. But here was the foundation for a car that he could own now. From their home in Inglewood, day after day Frank and his brother made the twelve-mile trek, carrying the T chassis back piece by piece.

Frank haunted the speed shops, and his favorite was Ray McDowell's in Hollywood. McDowell, who later won local fame as a builder of racing engines and components, gave Frank a junk Model T engine. He got it home and rebuilt it on his mother's kitchen floor. She said:

> Our house wasn't like other people's. That's how he started. McDowell was really a father to my Frank. He finally told him, "Bring all your plunder over to my place." He had a pit, and they got that Ford so it would go. That was his first car. He raced it at Ascot and it was hardly ever that anyone passed him. What little money he made just fed the car. He'd come home with burned feet and so tired that he couldn't get his shoes off. But that was what he had to do. He lived on grease and iron.

He lived to go fast. He never touched a drink or smoked or swore, not for any known moral reasons but because he considered such things wasteful and pointless. He went with just one girl in his life: the one he married. In his human relationships he was almost grotesquely stingy, but money meant nothing to him when there was a job to be done. That's what money was for.

He was no sentimentalist, as his mother knew best:

> One day in 1922 he came home and announced that he had to have five new racing tires for the Ford. I reminded him that we didn't have the $125. We hardly had a cent. He said, "Well, the furniture is paid for and you own some equity in the house—borrow the money!" I was quite sick at the time but Frank wasn't a boy that you could say no to. I went to the loan office and the manager agreed to put up the money but he insisted on talking to Frank "about the tires for this race car that he was going to break his neck in." He told Frank that he had to be paid back. Frank just said that if he made any money he would pay; he would not commit himself. The manager said, "That money is going to be paid back and it's going to be paid by you. I've seen mothers thrown in the gutter before by their kids." That's how Frank got his first set of racing tires.

Out of high school Frank continued his education at a technical night school and out of books at home. He was shy, quiet, and in spite of his obsession with personal goals, well liked. His mother went on:

> Everybody loved Frank . . . for his intelligence. He didn't have idle talk in him. When he talked, you listened, because you'd never heard it before. Maybe you've known someone like that. They're rare.

Somehow Frank came to the attention of the California Institute of Technology and was invited to take entrance examinations. He did and his mother was called to the office of Nobel Prize-winning physicist Robert Milliken. "Frank has the mind of a born scientist," he said. "Scholarships are scarce, but can't you find someone to back his education?"

"We don't know anyone like that," she answered. "But I could do all your family's sewing."

Frank did not attend Caltech.

Instead, he kept driving at Ascot and other small West Coast tracks. Once in a while someone would beat him but never when there was any good money at stake. This despite the fact that the whole car was built out of begged, borrowed, and otherwise scrounged parts. He even whittled his radiator shell out of wood and coated it with aluminum paint. Ernie Olson studied the local races faithfully and critically:

> No one ever sat in a race car like Frank. On a mile dirt track he seemed to begin his slide in the middle of the straightaway. Nobody ever imitated him. His skill and daring were tremendous. He wasn't exactly cool. He always got real bad butterflies before a race. He'd even vomit. In the car he'd say, "Pat me on the

First visit to Indianapolis: winner of the race. IMS

> shoulder." You'd do that and it seemed to fix the butterflies. But once the flag fell he was in command of everything. I once asked him what he thought about when he was racing. "All I think of from one second to the next," he said, "is how to drive to win."

When Frank wandered back to Indianapolis in 1926 just to watch the show he looked up the one man he knew there, Olson. Olson was in charge of the Miller of Bennett Hill and it was arranged for Frank to take a few practice laps on this course which he never had seen before. The twenty-three-year-old dirt-tracked the bricks, taking the turns in full, controlled slides, to the terror of the watchers on the outside rail.

"My God," said Hill. "That punk's getting around faster than I do."

When Frank came in Olson took him aside and asked how he liked the ride. Frank said that he liked it fine although he hadn't really been hard enough on the throttle to tell. Then Olson asked him how fast he thought he could get around the big oval if he tried. He allowed as how he should be able to knock four seconds off his time with no effort. That would have been a new track record and Olson said, "Kid, if Benny needs relief you've got a ride in the race."

But the day before the race Miller driver Pete Kreis was taken ill. Others had noted Lockhart's style, and he was offered the car. The rest is history:

> The race was a quarter over and the name Lockhart burned every wire when a fine rain swept the brick saucer clean of dirt and in its stead deposited a slippery, slimy surface that drove veteran after veteran into the pits. The pace slowed—for all except Lockhart, who semed to drive his foot farther into the floorboards.

When the race was called at 400 miles because of the rain the unknown kid from the West was five miles in front of his nearest competition. It was one of the most dramatic victories ever seen at the Speedway, and Frank was the talent discovery of that or any year. Harry Miller came to him and offered him a car for the rest of the season. This was praise from Caesar, but Frank would not accept this fulfillment of his ambition until he had the promise of the full-time help of the best racing mechanic he knew: Olson.

They made a nearly invincible team. Olson was a brainy practical engineer in his own right as well as a master mechanic, race strategist, and team manager. But Lockhart needed him not so much for his capacity to direct as for his ability to keep pace with Frank's ideas and to help in translating them into action. For Olson it was an unforgettable privilege to be a principal figure in the revolution that Lockhart wrought in American racing. Olson was only too glad to play the secondary role to such a man:

> When any part failed or gave trouble Frank would say, "There's a reason for this. Let's find out what it is." He might sit and study the problem for three or four days before doing a thing. But then he would have the answer.

Valves broke regularly in the racing engines of the day. Frank figured out why and eliminated this point of failure. Everyone used differentials on paved tracks until

Frank started winning with a locked rear end, which everyone imitated. Everyone was content with torque tubes until Lockhart discovered that the rear-axle housings still were bending under load. Then everyone used the radius rods which he introduced. These are only a few of his innovations. He experimented constantly and carried an elaborate machine shop with him from track to track, from coast to coast. What Miller and Duesenberg considered to be their best he refined to a spectacular degree. Soon, to have a Lockhart-type supercharger or connecting rod or almost any any other part was to be in the vanguard of the sport. As for surpassing the leader, or even catching up with him, this seemed not to be thought of.

Lockhart has been credited with the invention of the supercharger intercooler as far as racing cars are concerned, and Olson swears that the idea was entirely Frank's own. If so, he had precisely the right helpers in the Weisel brothers. He had a sharp eye for talent and hired the best he could find.

While working on his car at Miller's plant in Los Angeles in the summer of 1926 Frank became acquainted with John Weisel, a young Caltech engineering student who spent his vacations in Miller's employ. When Frank tried to interest Miller in detail improvements on his masterpieces the Old Man was scornful. So Frank hired Weisel to help him with design and drafting problems and through John met his brother Zenas, who recently had obtained his degree in mechanical engineering at Berkeley. His master's thesis had been a study of the supercharging of aircraft engines. Both brothers achieved distinguished records in the aircraft industry.

Frank had the excellent idea of building a power takeoff that could be bolted to the top of the transmission of his Buick road car and could be used for testing superchargers at racing speeds, anywhere and at any time. Zenas proposed a refinement: design a dynamometer into the device in order to be able to tell with precision the effect of any change in blower setup. During the first test, with the engine of Frank's Buick ticking over at 1,450 RPM and, through the gearbox, driving the blower at the

His first Miller. His wife was part of the team—naturally. GB COLLECTION

The fantastic International Class F Record at Muroc Dry Lake: 171 MPH one way. GB COLLECTION

equivalent of 7,500 crankshaft revolutions, they reached toward the blast from the blower outlet. From 18 inches away it was like touching a red-hot bar.

"I can design," said Weisel, "an air-cooled manifold that will get rid of an awful lot of that heat. I know it, and know what it can do. If it doesn't give you another eight MPH you won't owe me a dime."

The first time out with the new manifold Lockhart's Miller went 8.5 MPH faster than it ever had before. That was at the Culver City board track on March 6, 1927. Frank took one warm-up lap, raised his hand to signal the starter that he was on his way, and turned an all-time closed-course record of 144.2 MPH.

No one knew about the intercooler. A bit of it could be seen below the frame rails but it was passed off as an oil radiator. The car's hood was kept locked and never was raised in the presence of any but the few members of the team. Frank drove the big-time circuit from coast to coast and dominated almost everywhere with lap records and over-all victories. In one afternoon at Cleveland, for example, he broke over 100 records for the dirt-track mile. He kept the intercooler secret for almost a year.

That season Lockhart and Olson took their much-modified Miller 91 to Muroc Dry Lake in the Mojave Desert of California. The car long since had been purchased from Miller, who would not tolerate design changes being made on *his* cars. There Lockhart ran for official straightaway records against the strict AAA clocks. Said a newspaper account:

> Shattering all records for the 91-inch class at 171.02 MPH—the fastest time made by any car in the world excepting the Daytona monster, thirty times larger. Lockhart's two-way average against a heavy cross-wind was 164.85 MPH. The new record, made by a midget with the power of a giant, is a startling achievement when compared with the records established during recent years by cars with many times its size.

After witnessing decades of land-speed record attempts at Bonneville, Art Pillsbury said:

> The most spectacular mile I ever saw in my life was that one of Lockhart's. The dry lake was small and to get that record with his little 91 he went in a huge arc on two wheels and came sliding onto the straightaway. He stormed around on that alkali dirt when the average driver wouldn't even think of driving. *He* was a race driver. He could drive anything, anywhere.

In 1927 Lockhart came within an ace of winning the Indianapolis "500" for the second year in a row. He held the lead for 300 miles and then a connecting rod broke. After that he designed his own indestructible but beautiful connecting rods for the Miller engine. He finished the season a well-to-do young man. Every penny of his winnings was ear-marked for reinvestment in racing machinery.

chapter 25

THE LOCKHART SAGA-II

ON APRIL 28, 1926, J. G. Parry Thomas boosted the World's Land Speed Record to 170.624 MPH. His car was based on the Leyland straight-eight chassis which he had designed shortly after the war. The single overhead camshaft of the 443-cubic-inch engine was driven by eccentric rods such as Miller had contemplated using for his original eight.

On February 4, 1927, Captain Malcolm Campbell raised Thomas' record, on the same Pendine Sands beach in England, to 174.224. The displacement of his Napier engine was 1,360 cubic inches. Then Major H. O. D. Segrave came to Daytona, and on March 29, 1927, raised the LSR to 203.79 MPH. His Sunbeam's displacement was 2,760 cubic inches and the car weighed approximately 8,000 pounds. The trend was clear enough: big speeds demanded big cars.

Then on April 11 of that same year Lockhart achieved his two-way 164 and one-way 171 MPH on Muroc Dry Lake under highly adverse conditions. That he had accomplished this with a stock-bodied 91-cubic-inch race car that weighed only 1,500 pounds was a near-miracle that any layman could appreciate. The full implications rocked the racing world, and Frank Lockhart was the most impressed of all. On a long course such as Daytona he could have reached much greater speed. With a little more displacement he could easily become the world's fastest man on wheels. He would build a car—a tiny thoroughbred—to do the job.

There was more than vanity involved. Lockhart had designed a revolutionary

The result: the beach car, the LSR challenger, the Stutz Black Hawk. GB COLLECTION

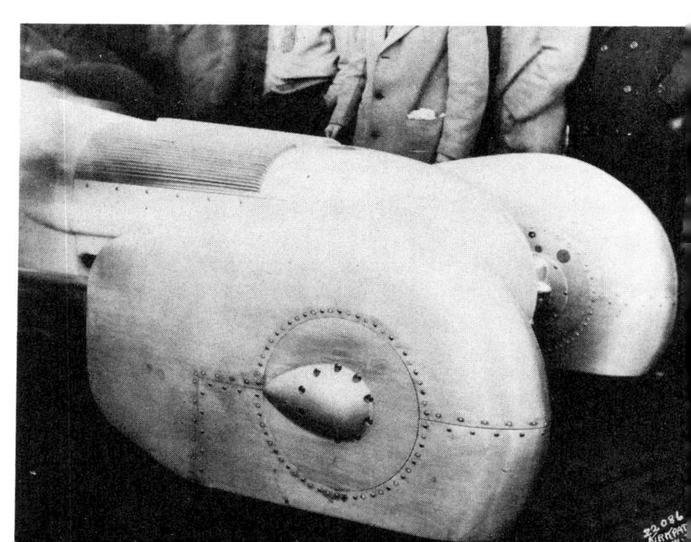

type of aircraft. It was to be powered by four engines and Lockhart stated that it would cross oceans and fly around the world. Gar Wood promised his complete support of the project, but it seemed so bizarre at the time that the press depicted Lockhart as a hopeless visionary. He needed to prove himself further in the eyes of the world. He began canvassing the automotive industry for backing for an all-American machine that would be as small and efficient as the new wave of British LSR locomotives was becoming gross. He approached Fred Moscovics, then president of the Stutz Motor Car Company. Moscovics said:

> Frank was the greatest natural mechanic I've ever come in contact with. He told me of his ambition to win the Land Speed Record for the United States. I knew that if anyone could do it, he could. I agreed to give him all the backing in my power. I got a group of wealthy sportsmen together who put up about $20,000. Frank agreed to call his machine the Stutz Black Hawk and I was able to add another $15,000 and put the facilities of the Stutz factory in Indianapolis at his disposal.

Olson in the meantime had been working hard to arrange Lockhart's participation in the Italian Grand Prix at Monza:

> I had the deal wrapped up when Frank told me he was going to build the beach car. I didn't exactly have a premonition of disaster. I just got physically ill. I just felt that it was the wrong thing to do and began to look for another job.

Lockhart wired the Weisel brothers, asking them to come to Indianapolis to work on the record-machine project for a guaranteed salary and for a minimum of four months. The challenge was fascinating to them. They hurried east and traveled around the country with him while Lockhart kept up with his racing commitments. Zenas recalled:

> My brother and I travelled with a suitcase each of clothes and a big trunk filled with books and instruments and a drawing board. We landed in Indianapolis, went with Frank to Altoona, then to Salem, where we got the car pretty well in mind, then to Buffalo, then to Detroit, then back to Indianapolis. Frank, being the star, had to stay in the best hotels. We always stayed at the YMCA. They were unusual working conditions. I can remember yet, him sitting in a chair in a hotel room while I measured his seated posture and laid out the frame kickups so they would just clear his elbows. We built the car around him.

Lockhart's original idea was to attack the record with two separate Lockhart-modified Miller 91's geared together after the fashion of Milton's LSR Duesenberg. He also visualized using a full envelope body. Zenas' engineering training had included heavy emphasis on aerodynamics as well as on engine design. In this case both were most intimately related, and he produced drawings of the sixteen cylinder Bugatti aircraft engine to show how a pair of straight eights could be joined in a very compact manner by integrating them into a common crankcase.

Concerned with the smallest possible frontal area for the car, Zenas squelched the envelope-body approach: it would sacrifice more than the meager horsepower available could afford. He held out for exposed wheels and for a torpedo shape that would enclose everything within the frame. Thus, not one square millimeter of frontal area would be given away, or allowed to absorb thrust that otherwise could be converted into speed.

He streamlined the axles, of course. He also streamlined the wheels that were sitting raggedly out in the air. He had listened to drivers who had run at high speeds and had been unhappy with the behavior of disc wheels. He reasoned that the effect of crosswinds on their steering was not due entirely to the lateral load created by the wind acting against the surface of the discs and pushing them in the direction of the wind. Zenas felt that the unexpectedly strong effects that had been noted were due, at least in part, to the fact that at high speeds the disc wheel, if inclined at an angle, became a lifting surface similar to an airplane wing. He also reasoned that the center of pressure on such a wheel was well ahead of the king pin, adding leverage to the tendency to steer in the direction of crosswinds. So Weisel designed the streamlined wheel-spats for the Black Hawk and made them long. He calculated their shape so that the center of pressure would be well aft of the king pins and would tend to counteract the car's tendency to steer in the direction of a crosswind and so that, acting as air rudders, they would contribute to the car's directional stability.

The spats were designed to be essentially parallel to the car's long axis at all times. Calculations indicated that, with an anticipated coefficient of friction on the beach of 0.5, the front wheels could not be turned more than zero degrees, six minutes at high speed without throwing the car into a skid. The lock-to-lock movement was

An only slightly distorted comparison of the Black Hawk against Seagrave's LSR-holding Sunbeam. *Autocar,* LONDON

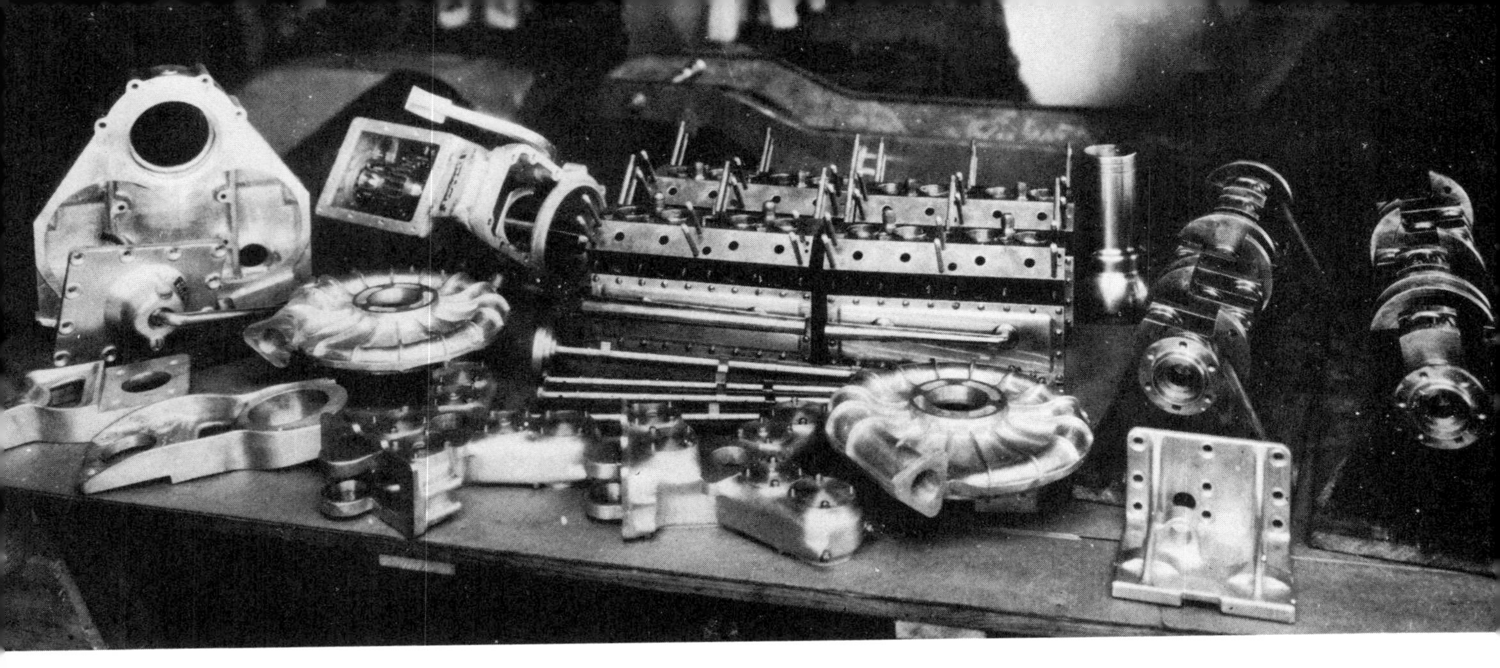

Components of the 16-cylinder engine. ZENAS WEISEL

The chassis. The delicate intercoolers formed part of the car's skin. ZENAS WEISEL

fixed just short of that figure. Total movement at the rim of the fourteen-inch steering wheel was just a quarter of an inch.

Efforts had been made in the past to give racing and record cars a streamlined appearance, at least, but their undersides generally had been left more or less uneven. And no thought was given to the internal streamlining of the engine space. This cost a tremendous price in wind drag which Weisel saved by enclosing the underside of the car and by sealing the engine compartment off from outside air, except for small openings for intake and exhaust. He originated the idea of eliminating the conventional radiator from straightaway machines. Coolant for the Black Hawk's power plant circulated through a tank which was filled with eighty pounds of ice.

The tiny, 181.6-cubic-inch engine was supercharged, of course, and it was equipped with intercoolers, each bank of cylinders having its own forced induction system. The intercooler on the Lockhart speedway car was mounted under the car's hood and its effectiveness depended upon the flow of air through the engine compartment. For the record machine Weisel designed an entirely new kind of intercooler: a delicate, almost filigreed aluminum casting which was contoured to serve as part of the car's hood.

The basic idea was to enclose all parts of the machine while still avoiding the obvious tactic of wrapping it in a full envelope. To get the body as small as possible while enclosing everything but the axles and wheels called for extremely narrow frame and for springs that also would be located within the body shell. Thus they could be placed only eleven inches apart.

Weisel wanted to secure the axles to the chassis without even a thousandth of an inch of mechanical play to possibly affect the car's steering. He designed and patented a unique suspension system. Each spring was a double cantilever machined from a solid steel billet. The front ends of the front springs were bolted to a steel spool which served as the center section of the front axle. The stub axles, in turn, were bolted to this assembly, outboard of the springs. The aft ends of the springs were bolted to bronze spacer blocks which were bolted to the frame. Total spring deflection was three-quarters of an inch, throughout which the king pins remained parallel with the frame.

The rear springs were of the same unusual construction; their front ends were bolted solidly to the frame and their rear ends pivoted around the rear axle housing. The worm-gear final drive was not dictated by Stutz' use of the same principle but by Stutz' reason for using it: it permitted lowering the car's height by about six inches.

The Black Hawk appears to have been the first car in the United States to benefit from extensive and serious use of wind-tunnel testing. The base-line for the the record machine was Frank's 91 speedway car and the knowledge that it had turned 171 MPH at 7,200 RPM. Therefore a precise scale model of the 91 was built, complete with radiator opening, dummy engine under the hood, hood louvers, knock-off hubs, dummy driver—all the features that contributed to wind drag. An equally precise model of the Black Hawk also was made. Utilizing the known top speed of the 91 and the data obtained by wind tunnel tests, the approximate available horsepower of both cars' engines could be extrapolated along with the Black Hawk's terminal velocity.

It was calculated that the completed car would weigh just under 2,800 pounds. This may sound light, but, for its size, this was a heavy machine. It also was calculated that the engine would deliver about 525 horsepower to the driving wheels at 7,500 RPM. Thorough wind-tunnel tests were conducted by the Curtiss Airplane Company, whose engineers translated the data into a top speed of 283 MPH.

The models then were sent to the Army wind tunnel at Wright Field for double-checking. There the top speed of the machine was calculated to be 281 MPH—excellent confirmation of the Curtiss figure. But by this time Lockhart, still racing in every championship event and with the brilliant talents of mechanic Jean Marcenac, had his 91 turning 8,200 RPM on the board track straightaways and going faster than ever. The horsepower estimate had to be revised upward, and final calculations credited the Black Hawk with a theoretical top speed of 330 MPH!

No word of this fantastic potential was allowed to leak beyond those persons most intimately concerned with the car. Even Moscovics in later years seemed not to know just how great its potential was. But Weisel always kept all of the original records and calculations.

Standard strategy among record breakers is not to exceed existing records by extravagant margins but, rather, to top them adequately, reap the rewards, rest on the laurels until a new record is established, and then, if the potential of the machine has not been exhausted, break the record again. Lockhart's plan was to boost the LSR to about 225 MPH, and most of his equipment tests were based on that simulated speed. Tire tests were among them.

Marcenac was in charge of the inner sanctum in the basement of the Stutz plant, where the Black Hawk was built. He helped Frank run the tire tests. The test rig consisted of a big Stutz straight-eight engine bolted to a concrete foundation and driving the shaft on which the wheels and tires were spun. Finally the time came for testing to destruction. The shop was cleared and the test tire, inflated to 125 PSI, was set spinning at the equivalent of 225 MPH. Marcenac and Lockhart hid behind concrete columns, and Frank fired a blast from a shotgun at the tire. The tremendous imbalance forces that suddenly were released tore the heavy engine from the block, and its aft end disintegrated. "Well," Lockhart said, "I'm done for if a tire blows."

He was depressed for days but never considered giving up the gamble.

In February of 1928, less than eight months after the preliminary paperwork on the Black Hawk had been completed, the car was finished and ready to run. That Malcolm Campbell had made his first trip to Daytona and on February 19 had raised the LSR to 206.956 MPH only made the Black Hawk seem more promising. The 35,000 dollars with which the project had been launched did not go far, and a large percentage of Lockhart's substantial racing winnings had been drained unexpectedly into the beach car. Estimates ran as high as 100,000 dollars. As Lockhart's investment in the project rose beyond all expectations his concern necessarily swung more and

Congratulations to British rival Malcolm Campbell. ZENAS WEISEL

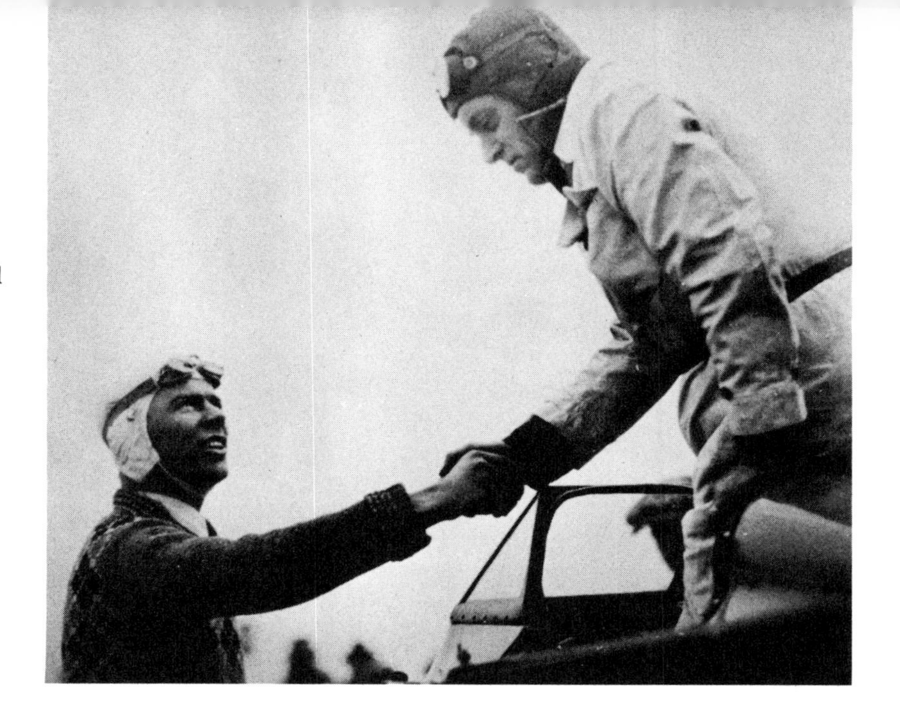

more from the pure challenge to the financial relief and reward that would come once the record was in his hands. Frank and his business manager, Bill Strum, would spend long hours computing the eventual take from track and vaudeville appearances, product endorsements and awards from manufacturers whose products he used.

Money became all-important to this extent: Lockhart always had driven on Firestone racing tires and had used them in his tests at the Stutz factory. But a new and short-lived tire manufacturer offered him 20,000 dollars if he set the new record using this firm's experimental high-speed tires. It was against Firestone policy to engage in such promotion, Harvey Firestone himself having laid down the dictum, "If our tires are good enough to risk your life on, they're good enough to buy." So Frank accepted the competition's offer.

Campbell was on the beach when the Black Hawk was shipped to Daytona in mid-February and practice runs began. Day after day the team struggled, but the car would not run above 180 MPH. Finally the source of the trouble was pinpointed: laminar flow across the supercharger inlets was starving the engine for air. Small scoops projecting into the air stream transformed the car's performance. But by this time the weather had become increasingly bad, and the AAA timers were ready to pick up and leave. They would stay just one more day: for Washington's Birthday and the already scheduled stock-car runs on the beach.

The weather was foul and the beach was miserable. But thousands of paid admissions had been taken in and, in spite of intermittent rain all day long, the crowds stayed to see if Lockhart would run. Finally, late in the afternoon, the little white car was towed to the south end of the course. The next thing the crowd was aware of was the 40,000 RPM scream of the Black Hawk's twin superchargers. Frank was on it all the way—moving at a good 225 the experts said—when he hit a patch of rain. Visibility was zero and the car edged into soft sand. Then a crash, two end-over-end flips, and a leap into the sea. The car landed on its belly, then skipped like a stone over the water, making slow rolls in the air. It came to rest wheels-down but nearly submerged in the surf.

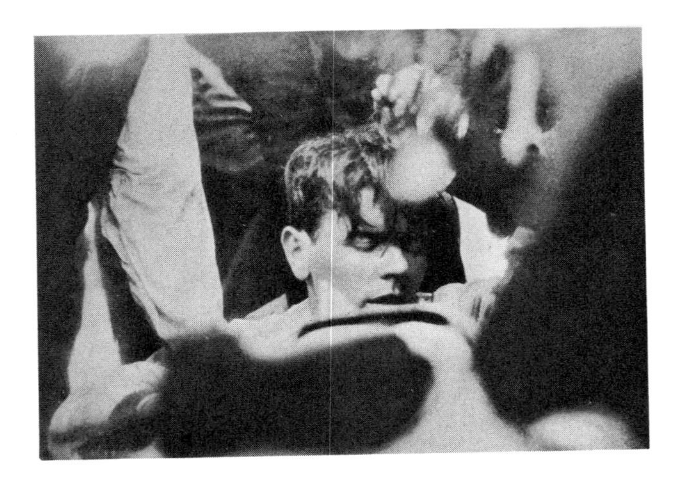

The first crash. GB COLLECTION

It seem impossible that Lockhart could be alive but after the first wave passed over the car he raised a hand, and the crowd, seemingly with one voice, roared, "Save him—he'll drown!" Moscovics was the first to rush into the surf and, holding Frank's head above the water, probably saved his life then and there. He was pinned in the body-tight cockpit but the car was dragged to shore. Aside from shock, augmented by Frank's irrational fear of water, his only injury was, miraculously, a couple of cut tendons in his right hand. Moscovics said:

> The car looked like a ruin but actually it wasn't badly damaged. We shipped it back to Indianapolis for repair. I did everything in the world to keep Frank from going back to the beach that year. But he had made no end of commitments and I guess he had to go.

In mid-April Lockhart and the Black Hawk were back at Daytona, and the eyes of the world were on them. He found a letter from his mother waiting for him:

> I told him that I was very ill and almost desperate. Would he send me just ten dollars? It would make all the difference in the world to me. A couple of days before the run I got a wire from him, the last I ever heard from him. He didn't mention the money. He just said, MA, I HAVE THE WORLD BY THE HORNS. YOU'LL NEVER HAVE TO PUSH A NEEDLE AGAIN. I'LL NEVER HAVE TO WORK ANYMORE. FRANK.

It was at dawn on April 25, 1928, that he made his first sally, a warm-up run to the south. The beach was perfect, and he returned north a little faster. Then south again with the superchargers' wail announcing that he was getting down to business. Then he began his fourth run, really flying. He was just about to enter the traps when a spray of sand shot from his right rear tire and the car came smashing, tumbling, thudding down the beach, and that was the end.

Officials who examined the evidence before the tide came in said that in braking at the end of his third run Frank had locked his rear wheels for nearly a hundred feet and that a clam shell had cut into the casing of the tire that blew.

Expert witnesses judged that Lockhart was moving at a good 220 MPH and still accelerating hard when the accident occurred. His previous run, the one that told him that the time had come for the big try, was clocked by the AAA at 198.29 MPH and still stands as the American National Class D (122- to 183-cubic-inch) record for the flying mile.

The last crash. GB COLLECTION

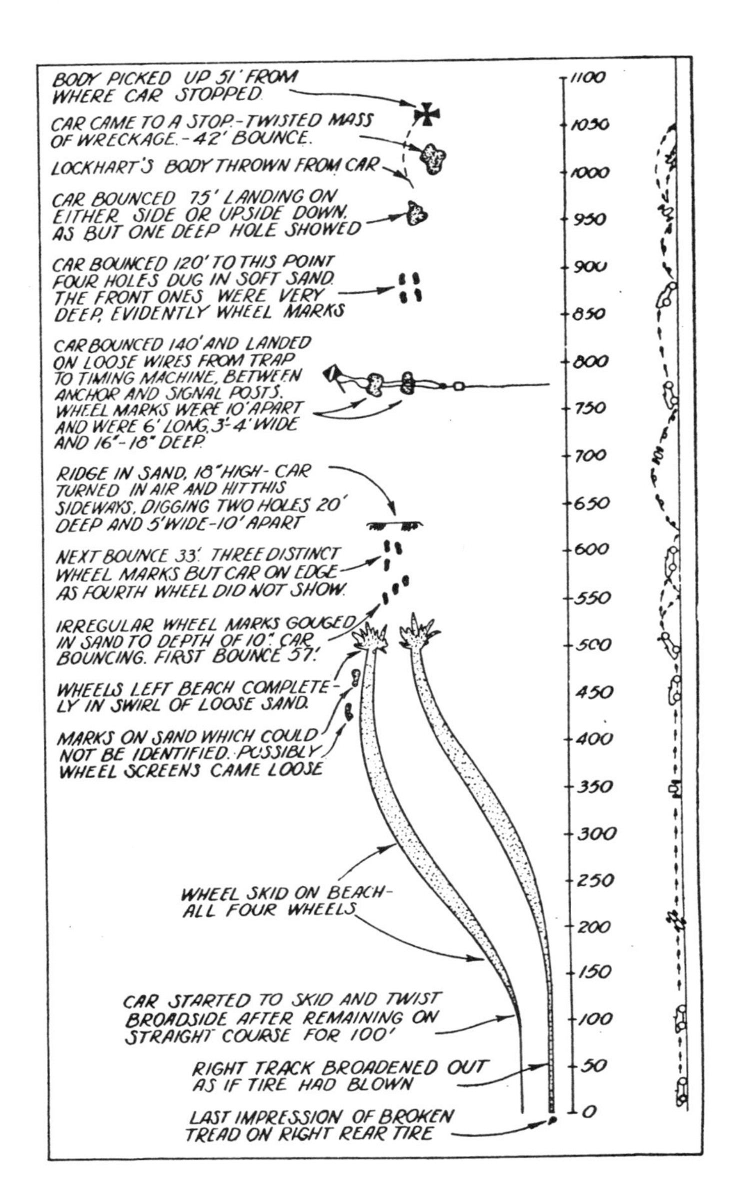

Chart of the last crash. GB COLLECTION

As we have seen, Lockhart's legacy lived on through his highly perfect machines. His two 91's remained virtually unbeatable until the end of the formula. Riley Brett bought the two-crankshaft V16 Black Hawk engine and put it away until 1939. It was twelve years old when it powered the Sampson Special to a new Indianapolis qualifying record of 129.431 MPH. Today it is enshrined in the Indianapolis Museum.

PART SIX

Project Time Machine

chapter 26

THE PACKARD CABLE SPECIALS

IN EARLY 1927 Leon Duray handed Harry Miller 15,000 dollars and became the owner of a new 91 front-drive car. It was gleaming black, with white numerals, frame rails, and wire wheels; Duray himself dressed entirely in black and was billed as The Black Devil. He was a hard man on machinery but as long as it held together he flew. In 1927 the absolute record for the 250-mile distance was Lockhart's 116.37 MPH. Duray kicked it up to 124.7. The following year he set the phenomenal 124.018 MPH Indianapolis (the course was all bricks and square corners then) lap record which stood until 1937. This was done with the help of his own unique intercooler and with this combination he went after the world's record for speed on a closed course. On June 14, 1928, on the still-unfinished Packard Proving Ground 2.5-mile banked concrete track he averaged 148.17 MPH. That record remained unbroken for twenty-six years.

The following year the Packard Cable Company (now a division of General Motors) sponsored a team of Miller 91's: the front-drives of Duray and Ralph Hepburn and the rear-drive of Tony Gulotta. Duray was the team manager, and the color scheme of the three cars was changed to rich purple, with bright yellow numerals, frame rails, and wheels. After the 1929 "500" Duray took the cars to Europe, ostensibly to compete in the Monza Grand Prix but also to create publicity for the new Cord car. It was being heavily promoted in Europe, where great stress was placed on its "front-

The Packard Cable Specials in the Bugatti factory in 1954. J. D. SCHEEL

drive, based on Miller patents." Duray was accompanied by Jean Marcenac, who had joined him after Lockhart's death, and he was shepherded in Europe by, of course, W. F. Bradley.

Duray's first essay on foreign soil was to attack the absolute closed-circuit record on the Montlhéry track, near Paris. Although its circumference was only half that of the Packard track he raised the five-mile record to 139.22 and the ten-mile record to 135.33 MPH. With Europe in the midst of a no-limit formula *Paris Match* eulogized:

> With a 1,500 cc car or, to be precise, with a car of 10 CV taxable horsepower, to achieve such speed proves clearly that the Americans are far in advance of us. We have no French cars of similar displacement capable of rivaling the speed of these Millers.

Or cars of any displacement, the writer might have added. These were galling words, and they galled Ettore Bugatti.

A month later Duray had his two cars at Monza. The course selected for the Grand Prix was a combination of the road course and the high-speed oval, 2.8 miles in total length. Giovanni Canestrini, dean of Italian automotive journalists, told me years later:

> Of course everyone knew that the American cars were totally unsuited to our kind of racing; their brakes and transmissions were rudimentary. So you can

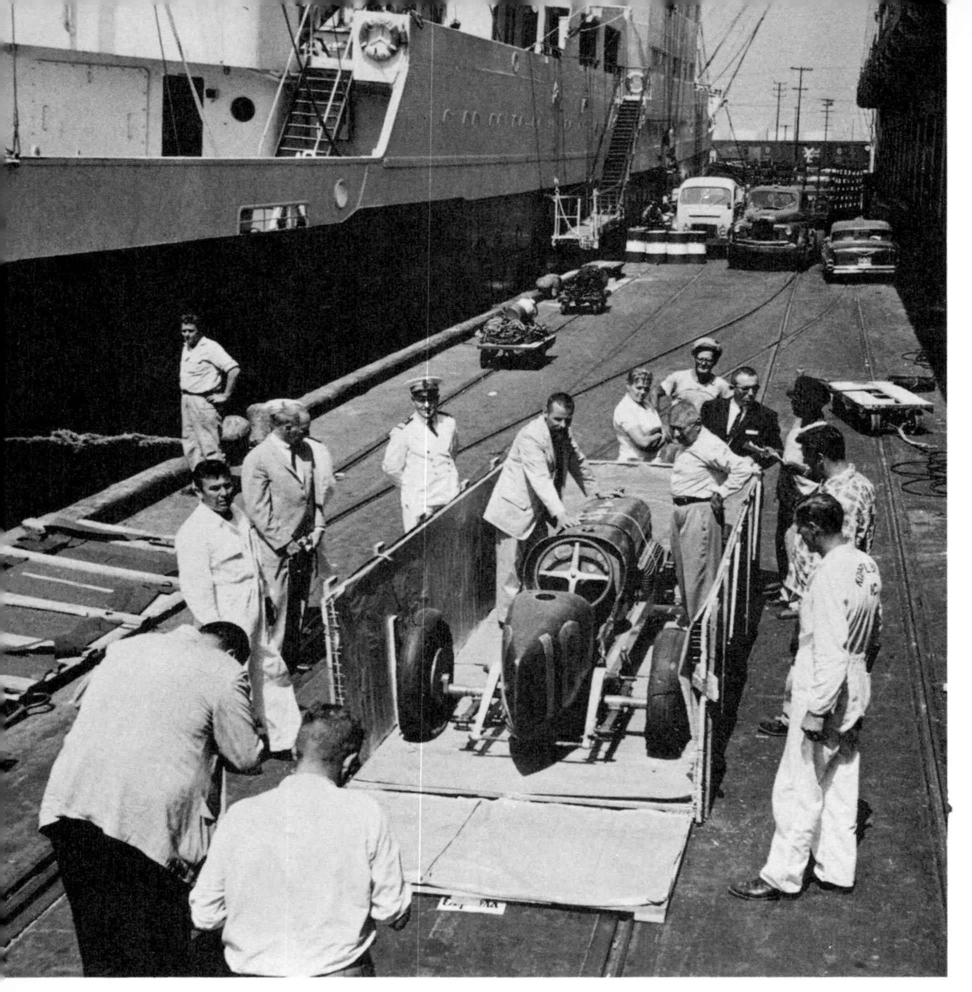

Historic moment. In the crate on the Los Angeles dock are Jean Marcenac and the author. ED ROTH

imagine our reaction when Duray took two—only two—practice laps and then demolished the lap record! His broadsliding style horrified the European drivers but he was a hero to the crowd.

The race was run in three eliminating heats and a final. Duray's competition in the first heat included Nuvolari and Arcangeli. But his 118.147 MPH was the fastest lap, and he led for the first three laps. Then, it was announced, his crankshaft bearings failed.

But Duray had a second car and was on the starting grid for the second heat, along with such talent as Varzi, Brilli Peri, Borzacchini, and Biondetti. This time he did not push his car as hard and diced with the leaders until Lap 12 when, again, he retired with alleged bearing failure.

When he returned to his pit Ettore Bugatti was waiting for him. He wanted the Millers and got them in exchange for a sum of cash and three new Bugatti 2.3 liter supercharged Targa Florio models. Duray took his acquisitions to Hollywood, where he operated a Bugatti agency just long enough to sell the cars. Bugatti took the Millers to Molsheim, copied their engines' top-end layout and introduced it on his Type 50; this was the turning point between the SOHC and DOHC Bugattis.

I first met Leon Duray in 1951. Among the many things we discussed before the death of that giant soon afterward was the thereabouts of the historic Packard Cable Specials. He assumed that they had vanished long ago.

In 1954 the story of a recent visit to Molsheim appeared in *Bugantics,* the publication of the Bugatti Owners Club of England. Included in the illustrations was a photo of the two ex-Duray machines, under a heavy coating of debris in the Bugatti factory. I promptly wrote to the author of the article and to the factory, asking the former for any information he could give and the latter if the cars were still at the factory, what condition they were in, and whether they could be purchased.

Author J. D. Scheel of the Royal Danish Embassy in London sent photos of the cars which he had taken in 1953. He had been informed that the cars were the Cooper Specials which Earl Cooper and Pete Kreis had driven in the Grand Prix of Europe in 1927, in which Cooper finished third. Evidently the cars had been covered with a protective coating of grease or cosmoline, and their Packard Cable Special identification was hidden.

I told many American collectors about the availability of the cars. Nothing happened. Finally, worried that they might be junked at any time, I decided to plunge. In October of 1958 and with Jean Marcenac as my interpreter I telephoned Pierre Marco of Bugatti and closed the deal.

The support that flooded in was most remarkable. I was on the editorial staff of *Car & Driver* magazine and, to my amazement, its owner, William B. Ziff, offered the loan of such funds as I might need. Both Firestone and Goodyear offered much-needed tires. The situation turned out to be fascinating to a great many people: bringing back to the States the only survivors of the Miller front-drive 91, and highly distinguished survivors at that. Offers of help from old friends and from friends I did not know existed came in. While the cars were being prepared for shipment, *C/D*

The author takes a plug reading while customs inspector Melvin Bowden makes sure the product is home-grown. LOS ANGELES *Herald-Express*

Ensconced in the museum.
GB COLLECTION

technical editor Karl Ludvigsen saw them in Molsheim and wrote:

> The cars are ready to go as they stand. Considering the time and the trials these two cars have been through their condition is literally staggering. The purple paint, the Packard Cable Special lettering, all is tatty but intact. It was a blinding glimpse into the past.

On July 30, 1959, the French Line's *MS Wyoming* docked in Los Angeles Harbor. Two big crates were swung onto the dock. TV cameras ground and NBC's *Monitor* taped the comments as Marcenac, Olson, and de Paolo shared with me the unforgettable thrill of seeing those battered, veteran machines rolled onto their home soil once more. The crew of the *Wyoming* was infected by the spirit of the Franco-American occasion. The skipper had the afterdeck decorated as though it were Bastile Day, and toasts were exchanged until sundown. Then Lindley Bothwell's transporter carried the Millers to my hideaway in the Santa Monica Mountains. It was just over the hill from Harry Miller's old ranch.

Then the work began; the hundreds of hours that had been invested up to that point were as nothing. But my first act was to get both machines legally registered in my name, as though they had been passenger cars.

Instinct told me to begin work on the ex-Hepburn car, which I dismantled down to the bare frame rails. And then I received a call from Karl Kizer, guiding spirit of the Indianapolis Speedway Museum. He wanted the Duray machine, which was so significant in Speedway history. This was precisely the future that I had hoped for this car, and I offered it to him at a non-profit price. The following day Tony Hulman called, accepting the offer. Soon the immortal Duray front-drive was on its way to Indianapolis, where it was restored to its gleaming, original running condition.

My own project had to be fitted into a tightly packed schedule of work and travel. Out of sheer jealously I would entrust only the most specialized tasks to outside hands. The loosening of each nut and bolt was an adventure. I found degrees of finish that were breath-taking, hidden within the Miller chassis since it had been assembled over three decades before. The experience was one long, long unveiling of a beloved's secrets. Among them, she told me why—"she" being the car, of course—Duray had had to wheel her into the pits at Monza. She was perfect throughout except for two things. Her front shock-absorber bolts were bent into a sharp "V" and were ready to shear. And her front-drive unit's finely shaped housing had cracked. Duray had gotten the gears to downshift but the case could not support the load.

(Left to right) The author, Leo Goossen, and Louis Meyer. WAYNE THOMS

The ex-Duray machine, turned over by the author to the Indianapolis Speedway Museum, after full restoration under Karl Kizer's direction. GB COLLECTION

I plugged away at the job, making progress at satisfying rate. Late in September, 1961, the Los Angeles County Museum begged for use of the car, to be a major feature in a special two-month exhibit devoted to the last fifty years of Southern California history. The deadline was November 1 and, approving of the idea, I devoted the entire month of October to completing the job. My car and the ex-Duray Indianapolis car appeared simultaneously in their pristine originality, Duray's in the Midwest and Hepburn's in the West. They were somewhat exquisite, these pieces of evidence of almost-forgotten glory.

After the big Los Angeles exhibit the ex-Hepburn car was shipped to the New York International Auto Show, where, in perhaps less provincial surroundings, it was a smashing attraction. Then it was returned, on loan, to the Los Angeles County Museum, where, over my most violent protests and in spite of my registered ownership, it remains out of my possession.

appendix I

AMERICAN BOARD TRACKS

Location	*Designer*	*Length, Miles*	*Years Active*
Playa del Rey, Calif.	JP	1.0	1910 - 1913
Elmhurst, Calif.	JP	0.5	1911 - 1913
Chicago, Ill. (Maywood)	UN	2.0	1915 - 1917
Des Moines, Iowa	JP	1.0	1915 - 1917
Omaha, Neb.	JP	1.25	1915 - 1917
Brooklyn, N. Y. (Sheepshead Bay)	UN	2.0	1915 - 1919
Uniontown, Penna.	JP	1.125	1916 - 1922
Cincinnati, Ohio	UN	2.0	1916 - 1919
Tacoma, Wash.	UN	2.0	1915 - 1921
Beverly Hills, Calif.	ACP	1.25	1920 - 1924
Fresno, Calif.	ACP & JP	1.0	1920 - 1927
San Carlos, Calif.	ACP & JP	1.25	1921 - 1922
Cotati, Calif.	JP	1.25	1921 - 1922
Kansas City, Mo.	ACP & JP	1.25	1922 - 1924
Altoona, Penna.	ACP & JP	1.25	1923 - 1931
Charlotte, N. C.	JP	1.25	1924 - 1926
Culver City, Calif.	ACP & JP	1.25	1924 - 1927
Salem, N. H. (Rockingham)	ACP & JP	1.25	1925 - 1927
Laurel, Md.	JP	1.125	1925 - 1926
Miami, Fla. (Fulford-by-the-Sea)	UN	1.25	1926 - 1927
Amatol, N. J. (Atlantic City)	ACP & JP	1.5	1926 - 1928
Woodbridge, N. J.	UN	0.5	1929 - 1931
Akron, Ohio	UN	0.5	
Bridgeville, Penna.	UN	0.5	

JP = Jack Prince   ACP = Arthur C. Pillsbury   UN = Unknown

appendix II

RESULTS OF MAJOR AMERICAN RACES - 1915 THROUGH 1929

BS = Board Speedway BRS = Brick Speedway CS = Concrete Speedway AS = Asphalt Speedway
DT = Dirt Track RC = Road Course

NOTE: The bulk of the following data was originally compiled by Charles L. Betts, Jr., distinguished engineer, author, historian, and editor.

Date	*Event*	*Car*	*Driver*	*Avg. Spd.* MPH
		1915		
Jan. 9	Point Loma Road Race, San Diego. RC. 305.082 miles	Stutz	E. Cooper	65.05
Feb. 3	Glendale Road Race, Glendale, Calif. RC. 103 miles	Duesenberg	E. O'Donnell	47.90
Feb. 27	Sixth American Grand Prix, San Francisco. RC. 400.28 miles	Peugeot	D. Resta	56.12
Mar. 6	Tenth Vanderbilt Cup Race, San Francisco. RC. 300.3 miles	Peugeot	D. Resta	67.50
Mar. 17	Venice Grand Prix, Venice, Calif. RC. 300 miles	Maxwell	B. Oldfield	68.80
May 1	SW Sweepstakes, Oklahoma City. RC. 199.532 miles	Peugeot	B. Burman	67.98

Date	Event	Car	Driver	Avg. Spd. MPH
May 30	Indianapolis BRS. 500 miles	Mercedes	R. de Palma	89.84
June 26	Chicago BS. 500 miles	Peugeot	D. Resta	97.58
July 3	Sioux City DT. 300 miles	Maxwell	E. Rickenbacker	74.70
July 5	Tacoma BS.			
	250 miles	Mercer	Ruckstall	84.50
	200 miles	Mercer	E. Pullen	85.20
	100 miles	Parsons	Parsons	79.50
	Omaha BS. 300 miles	Maxwell	E. Rickenbacker	91.74
Aug. 7	Chicago BS. 100 miles	Peugeot	D. Resta	101.86
	Des Moines BS. 300 miles	Duesenberg	R. Mulford	87.00
Aug. 20	Elgin Road Race. RC. 301 miles	Stutz	G. Anderson	77.26
Sept. 4	Minneapolis 2-mile CS. 500 miles	Stutz	E. Cooper	86.35
Sept. 18	Providence, R.I. 1-mile AS.			
	100 miles	Maxwell	E. Rickenbacker	67.10
	25 miles	Peugeot	B. Burman	69.76
Oct. 9	Sheepshead Bay BS. 350 miles	Stutz	G. Anderson	102.60
Nov. 2	Sheepshead Bay BS. 100 miles	Peugeot	D. Resta	105.39
	1916			
Apr. 8	Corona Road Race, Corona, Calif. RC. 300 miles	Duesenberg	E. O'Donnell	85.60
Apr. 15	Ascot 1-mile DT, Los Angeles. 150 miles	Duesenberg	E. O'Donnell	65.40
May 13	Sheepshead Bay BS.			
	150 miles	Maxwell	E. Rickenbacker	96.23
	50 miles	Peugeot	R. Mulford	104.34
	20 miles	Peugeot	J. Aitken	106.71
May 30	Indianapolis BRS. 300 miles	Peugeot	D. Resta	86.23
June 10	Chicago BS. 300 miles	Peugeot	D. Resta	98.70
June 24	Des Moines BS. 150 miles	Mercedes	R. de Palma	92.66
July 4	Minneapolis CS. 150 miles	Mercedes	R. de Palma	90.80
July 15	Omaha BS. 150 miles	Peugeot	D. Resta	99.02
	50 miles	Mercedes	R. de Palma	103.45
July 22	Kansas City 1.125-mile DT. 100 miles	Mercedes	R. de Palma	58.40
Aug. 5	Tacoma BS. 300 miles	Maxwell	E. Rickenbacker	89.30

Date	*Event*	*Car*	*Driver*	*Avg. Spd.* MPH
Sept. 4	Cincinnati BS. 300 miles	Peugeot	J. Aitken	97.06
Sept. 9	Indianapolis BRS. 20 miles	Peugeot	J. Aitken	95.00
	50 miles	Peugeot	J. Aitken	91.50
	100 miles	Peugeot	J. Aitken	89.04
Sept. 30	Sheepshead Bay BS. 250 miles	Peugeot	J. Aitken	104.83
Oct. 14	Chicago BS. 250 miles	Peugeot	D. Resta	103.99
Oct. 30	Sheepshead Bay BS. 100 miles	Peugeot	J. Aitken	105.95
	50 miles	Delage	J. LeCain	104.20
Nov. 16	Vanderbilt Cup Race, Santa Monica. RC. 294.03 miles	Peugeot	D. Resta	86.98
Nov. 18	Seventh American Grand Prix, Santa Monica. RC. 403.24 miles	Peugeot	H. Wilcox	85.59
Nov. 30	Ascot AS. 150 miles	Duesenberg	E. Rickenbacker	65.00
Dec. 2	Uniontown BS. 112.5 miles	Frontenac	L. Chevrolet	102.00
		1917		
May 30	Cincinnati BS. 250 miles	Frontenac	L. Chevrolet	102.18
June 16	Chicago BS. 250 miles	Stutz	E. Cooper	103.15
July 4	Omaha BS. 150 miles	Hudson	R. Mulford	101.40
	50 miles	Hoskins	D. Lewis	102.85
	Uniontown BS. 112 miles	Peerless	I. Fetterman	90.90
Aug. 9	St. Louis 1-mile DT.			
	Records: 1 mile	Miller	B. Oldfield	80.00
	50 miles	Miller	B. Oldfield	73.50
Sept. 22	Sheepshead Bay BS. 100 miles	Frontenac	L. Chevrolet	110.40
	10 miles	Delage	J. LeCain	97.80
		1918		
June 1	Sheepshead Bay BS. 100 miles	Packard	R. de Palma	102.00
June 26	Chicago BS. 100 miles	Frontenac	L. Chevrolet	108.12
		1919		
Mar. 15	Santa Monica Road Race. RC. 250.24 miles	"Chevrolet"	C. Durant	81.6
Mar. 23	Ascot AS. 150 miles	Roamer	R. Sarles	71.0
May 20	Uniontown BS. 112.5 miles	Duesenberg	T. Milton	96.2
May 30	Indianapolis BRS. 500 miles	Peugeot	H. Wilcox	88.1

Date	Event	Car	Driver	Avg. Spd. MPH
June 14	Sheepshead Bay BS. 50 miles	Packard	R. de Palma	113.8
July 4	Sheepshead Bay BS. 100 miles	Frontenac	G. Chevrolet	110.5
	Tacoma BS. 80 miles	Frontenac	L. Chevrolet	100.0
Aug. 23	Elgin Road Race. RC. 301 miles	Duesenberg	T. Milton	73.9
Sept. 1	Uniontown BS. 112.5 miles	Frontenac	J. Boyer	96.4
Sept. 20	Sheepshead Bay BS. 150 miles	Frontenac	G. Chevrolet	109.5
		1920		
Feb. 29	Beverly Hills BS. 250 miles	Duesenberg	J. Murphy	103.0
May 30	Indianapolis BRS. 500 miles	Monroe	G. Chevrolet	88.5
June 19	Uniontown BS. 225 miles	Duesenberg	T. Milton	94.9
Aug. 28	Elgin Road Race. RC. 251 miles	Ballot	R. de Palma	79.0
Oct. –	Fresno, Calif. BS. 250 miles	Duesenberg	J. Murphy	–
		1921		
Feb. 28	Beverly Hills BS. 50 miles	Ballot	R. de Palma	107.3
May 30	Indianapolis BRS. 500 miles	Frontenac	T. Milton	89.6
July 4	Tacoma BS. 250 miles	Miller	T. Milton	96.9
Aug. 15	Cotati, Calif. BS. 150 miles	Duesenberg	E. Hearne	110.2
Sept. 9	Indianapolis BRS. 50 miles	Frontenac	H. Wilcox	97.5
Nov. 30	Beverly Hills BS. 250 miles	Duesenberg	E. Hearne	109.7
		1922		
Mar. 5	Beverly Hills BS. 250 miles	Duesenberg	T. Milton	110.8
Apr. 27	Fresno BS. –	Miller-Dues	J. Murphy	103.0
May 7	Cotati BS. 100 miles	Miller-Dues	J. Murphy	114.2
	50 miles	Fiat	P. Bordino	114.5
May 30	Indianapolis BRS. 500 miles	Miller-Dues	J. Murphy	94.5
Aug. 7	Cotati BS. 150 miles	Miller	B. Hill	114.0
	100 miles	Miller	F. Elliott	113.7
	50 miles	Miller	F. Elliott	107.5
Sept. 17	Kansas City BS. 300 miles	Miller	T. Milton	107.0
Oct. 2	Fresno BS. 150 miles	Miller	B. Hill	102.0
Oct. 30	Beverly Hills BS. 100 miles	Miller	B. Hill	114.5
Nov. –	Beverly Hills BS. 250 miles	Miller	J. Murphy	114.6

Date	*Event*	*Car*	*Driver*	*Avg. Spd.* MPH
		1923		
Feb. 22	Beverly Hills BS. 250 miles	Miller	J. Murphy	115.8
Apr. 30	Fresno BS. 150 miles	Miller	J. Murphy	103.0
May 30	Indianapolis BRS. 500 miles	Miller	T. Milton	90.9
July 4	Kansas City BS. 250 miles	Miller	E. Hearne	105.8
Sept. 4	Altoona BS. 200 miles	Miller	E. Hearne	111.5
Oct. 1	Fresno BS. 150 miles	Miller	H. Hartz	103.6
Oct. 21	Kansas City BS. 250 miles	Miller	H. Fengler	113.4
Nov. 29	Beverly Hills BS. 250 miles	Miller	B. Hill	112.0
Dec. —	Syracuse 1-mile DT. 100 miles	Miller	T. Milton	80.0
		1924		
Feb. 24	Beverly Hills BS. 250 miles	Miller	H. Fengler	116.6
May 30	Indianapolis BRS. 500 miles	Duesenberg	Corum-Boyer	98.2
June 14	Altoona BS. 250 miles	Miller	J. Murphy	114.7
July 4	Kansas City BS. 150 miles	Miller	J. Murphy	114.4
Sept. 1	Altoona BS. 250 miles	Miller	J. Murphy	113.9
Sept. 15	Syracuse DT. 150 miles	Duesenberg	P. Shafer	70.1
Oct. 27	Charlotte BS. 250 miles	Miller	T. Milton	118.2
Oct. —	Fresno BS. 150 miles	Miller	E. Cooper	105.0
Dec. 15	Culver City BS. 250 miles	Miller	B. Hill	126.9
		1925		
Mar. 4	Culver City BS. 250 miles	Miller	T. Milton	126.9
Apr. 19	Culver City BS. 25 miles	Miller	L. Duray	134.0
	25 miles	Duesenberg	P. de Paolo	135.0
	25 miles	Fiat	P. Bordino	133.0
	25 miles	Miller	R. McDonough	130.0
	50 miles	Miller	H. Hartz	135.2
Apr. 30	Fresno BS. 150 miles	Duesenberg	P. de Paolo	—
May 11	Charlotte BS. 250 miles	Miller	E. Cooper	121.6
May 30	Indianapolis BRS. 500 miles	Duesenberg	P. de Paolo	101.1
June 13	Altoona BS. 250 miles	Duesenberg	P. de Paolo	115.9
July 4	Salem, N. H. 1-mile DT.	— —	R. de Palma	76.9
July 11	Laurel BS. 250 miles	Duesenberg	P. de Paolo	123.3

Date	Event	Car	Driver	Avg. Spd. MPH
Oct. 24	Laurel BS. 250 miles	Miller	R. McDonough	126.3
Oct. 31	Salem BS. 250 miles	Duesenberg	P. de Paolo	125.2
		1926		
Feb. 22	Miami BS. 300 miles	Duesenberg	P. de Paolo	129.3
Mar. 21	Culver City BS. — 122 cu. in. cars only. 250 miles	Miller	B. Hill	131.3
	10 miles	Miller	L. Duray	136.0
May 1	Atlantic City BS. (122's only) 300 miles	Miller	H. Hartz	135.2
May 30	Indianapolis BRS. (91's only) 500 miles, stopped at 400	Miller	F. Lockhart	95.9
June 12	Altoona BS. 250 miles	Miller	D. Lewis	112.4
June 17	Philadelphia 1-mile DT. 50 miles	Duesenberg	F. Winnai	70.0
June 19	Laurel BS. 100 miles	Duesenberg	J. Gleason	105.0
July 4	Philadelphia DT. 50 miles	Miller	R. Keech	69.8
July 17	Atlantic City BS. 60 miles	Miller	N. Batten	120.8
	60 miles	Miller	F. Comer	124.7
	60 miles	Miller	H. Hartz	128.7
	120 miles	Miller	H. Hartz	123.4
July —	Salem BS. 50 miles	— —	P. de Paolo	128.3
	200 miles	Miller	E. Cooper	—
Aug. 7	Philadelphia DT. 25 miles	Duesenberg	F. Winnai	76.8
	1-mile DT. record	Duesenberg	F. Winnai	90.2
Aug. 23	Charlotte BS. 25 miles	Miller	E. Cooper	128.9
	25 miles	Miller	D. Lewis	125.2
	150 miles	Miller	F. Lockhart	120.5
Sept. 18	Altoona BS. 250 miles	Miller	F. Lockhart	117.0
Nov. 11	Charlotte BS. 25 miles	Miller	F. Lockhart	132.4
	25 miles	Miller	D. Lewis	127.0
		1927		
Mar. 6	Culver City BS. 250 miles	Miller	L. Duray	124.7
May 7	Atlantic City BS. 200 miles	Miller	D. Lewis	130.6
	Lap record	Miller	F. Lockhart	147.7
May 30	Indianapolis BRS. 500 miles	Duesenberg	G. Souders	97.5
June 11	Altoona BS. 200 miles	Duesenberg	P. de Paolo	116.6

Date	Event	Car	Driver	Avg. Spd. MPH
July 4	Salem BS. 200 miles	Miller	P. de Paolo	124.3
Sept. 5	Altoona BS. 250 miles	Miller	F. Lockhart	117.5
Sept. 25	Cleveland 1-mile DT. 1-mile DT. record	Miller	F. Lockhart	92.5
Oct. 12	Salem BS. 75 miles	Miller	F. Lockhart	126.7
		1928		
May 30	Indianapolis BRS. 500 miles	Miller	L. Meyer	99.5
June 8	Detroit 1-mile DT. 100 miles	Miller	R. Keech	77.9
July 4	Atlantic City BS. 100 miles	Duesenberg	F. Winnai	101.0
July 4	Salem BS. 185 miles	Miller	R. Keech	—
Aug. 19	Altoona BS. 200 miles	Miller	L. Meyer	118.0
Sept. 16	Atlantic City BS. 100 miles	Miller	R. Keech	131.6
		1929		
May 26	Toledo 1-mile DT. 100 miles	Miller	W. Shaw	76.46
May 30	Indianapolis BRS. 500 miles	Miller	R. Keech	97.59
	Bridgeville, Pa. 1-mile DT. 100 miles	Miller	W. Shaw	67.30
June 2	Cleveland 1-mile DT. 85 miles	Miller	W. Shaw	71.57
June 9	Detroit 1-mile DT. 100 miles	Miller	C. Woodbury	76.22
June 15	Altoona BS. 200 miles. stopped at 151	Miller	R. Keech (post-humously)	118.4
June 30	Woodbridge BS. 100 miles	Miller	L. Moore	74.22
July 4	Bridgeville 1-mile DT. 100 miles	Miller	W. Shaw	71.33
Aug. 18	Woodbridge BS. 100 miles	Miller	L. Moore	73.0
	Toledo 1-mile DT. 100 miles	Miller	W. Shaw	78.94
Aug. 31	Syracuse 1-mile DT. 100 miles	Miller	W. Shaw	81.07
Sept. 2	Altoona BS. 200 miles	Miller	L. Meyer	112.0

appendix III

RECORDS SET BY AMERICAN 91-CU.-IN. RACING CARS

Distance, miles	*Car*	*Driver*	*Speed,* MPH	*Course*
1	Miller	Lockhart	164.0	Muroc Dry Lake
1.5	Miller	Lockhart	147.7	Atlantic City BS
2.5	Miller	Duray	148.1	Utica, Mich. CS
5	Miller	Duray	139.2	Montlhéry, France
10	Miller	Lockhart	135.6	Atlantic City BS
25	Miller	Lockhart	132.4	Charlotte BS
50	Miller	Lewis	131.0	Atlantic City BS
100	Miller	Keech	131.8	Atlantic City BS
150	Miller	Woodbury	132.0	Atlantic City BS
200	Miller	Lewis	130.0	Atlantic City BS
250	Miller	Duray	124.7	Culver City BS
300	Duesenberg	Gleason	103.2	Indianapolis BRS
350	Duesenberg	Gleason	102.0	Indianapolis BRS
400	Miller	Gulotta	100.9	Indianapolis BRS
500	Miller	Meyer	99.4	Indianapolis BRS

appendix IV

ENGINE SPECIFICATIONS

Make	Year Intro.	Cyls.	Bore x Stroke		Displacement		Cam-shafts
			Ins.	Mm.	Cu. In.	Cc.	
DUESENBERG	1912	4	4.316 x 6.0	109 x 152	351	5750	1
	1914	4	4.375 x 6.0	111 x 152	361	5920	1
	1916	4	3.9 x 6.0	99 x 152	299	4900	1
	1916	4	3.75 x 6.75	95 x 171	298	4890	2
	1916	V12	4.875 x 7.0	124 x 178	1568	25,700	1
	1918	V16	6.0 x 7.5	152 x 190	3393	55,600	1
	1919	8	2.875 x 5.0	73 x 127	260	4260	1
	1919	8	3.0 x 5.25	76 x 133	296	4850	SOHC
	1920	8	2.50 x 4.625	63 x 117	182	2980	SOHC
	1922	8	2.50 x 4.625	63 x 117	182	2980	DOHC
	1923	8	2.375 x 3.422	60 x 87	121	1990	DOHC
	1924	8	2.375 x 3.422	60 x 87	121	1990	DOHC
	1926	8	2.286 x 2.75	58 x 70	90	1480	DOHC
	1926	8	2.286 x 2.75	58 x 70	90	1480	—
	1927	8	2.1875 x 3.0	55 x 76	90	1485	DOHC

Valves/ Cyl.	*Valve Actuation*	*Ignition*	*Spark Plugs/ Cyl.*	*Main Bearings*	*Crankcase Type*	*Notes*
2	Walking Beam	Magneto	2	2	Barrel	Mason-Maytag. About 95 HP.
2			2	2		First "Duesenberg."
2			2	2		Speedway Engine.
4			2	2		Training-Plane Engine Variant.
2			2	4	Split	Aero—300 HP @ 1400 RPM.
3	Modified W. B.		2	5	Barrel	Aero—700-800 HP @ 1250 RPM.
2	W. B.	Battery	1	3	Split	Walking-Beam Touring Engine 100 HP @ 3000 RPM.
3	Rockers		2	3	Barrel	First Racing Eight.
3			1	3		"Grand Prix" Straight Eight.
4	Cups		1	3		125 HP @ 4000 RPM.
4	Cups		1	3 & 5		Shaft & Bevel Cam Drive Through Indy., 1923. Gear Train Thereafter.
2	Cups		1	3		First Success With Supercharging.
2	Cups		1	5		Magneto Ignition Used on a Few Engines.
—	—		1	5		Rotary Intake Valves, Window-Type Exhaust Valves.
2	Cups		1	5		160 HP in 1927.

Make	Year Intro.	Cyls.	Bore x Stroke		Displacement		Cam-shafts
			Ins.	Mm.	Cu. In.	Cc.	
FRONTENAC	1916	4	3.870 x 6.375	98 x 162	299	4900	SOHC
	1920	4	3.125 x 5.9375	79 x 151	182	2980	DOHC
	1921	8	2.625 x 4.21875	67 x 107	182	2980	DOHC
	1922	4	—	—	182	2980	DOHC
	1923	4	3.0 x 4.3125	76 x 110	122	1995	—
	1923	4	3.115 x 4.0	79 x 104	182	2980	DOHC
MAXWELL	1914	4	3.75 x 6.75	95 x 171	298	4885	SOHC
MILLER	1916	4	3.625 x 7.0	92 x 178	289	4740	SOHC
	1917	V12	5.0 x 6.0	127 x 152	1414	23,200	1
	1919	4	3.09375 x 6.0	79 x 152	182	2980	SOHC
	1921	8	2.6875 x 4.0	68 x 104	182	2980	DOHC
	1923	8	2.375 x 3.5	60 x 89	121	1980	DOHC
	1926	8	2.187 x 3.0	56 x 76	90	1478	DOHC
	1926	4	3.406 x 4.125	87 x 105	150	2458	DOHC
PREMIER	1916	4	3.66 x 6.625	93 x 168	275	4500	DOHC

Valves/ Cyl.	*Valve Actuation*	*Ignition*	*Spark Plugs/ Cyl.*	*Main Bearings*	*Crankcase Type*	*Notes*
4	R R	Magneto	1	3	Split	130-140 HP—Shaft & Bevel Cam Drive. "Aluminum" Engine.
4	Fingers		1	3	Barrel	Also Monroe. Delco Distributor on Winning Car.
4	Cups	Battery	2	5	Barrel	Spur-Gear Cam Drive.
4	Cups		1	3	Barrel	Four-Cyl. Version of Straight Eight.
—	—		1	3	Barrel	Scheel-Frontenac. Rotary Intake, Window Exhaust.
2	Mushroom	Magneto	2	3	Split	Fronty-Ford. Chain Cam Drive. 80 HP "SR" Head.
2 & 4	Rockers		1	3		Designed by Ray Harroun.
4	Rockers	Magneto	1	3	Barrel	125 HP @ 2950 RPM. Desmodromic Valves (Experimentally).
2	Modified W. B.		1	5	Split	Aero—500 HP.
4	Rockers		1	3	Barrel	Experimental Hydraulic Brakes.
4	Cups		1	3	Barrel	125 HP @ 4000 RPM, Battery Ignition on First Engines.
2	Cups		1	5	Barrel	120 HP @ 4500-5000 RPM, 203 HP Supercharged @ 5800.
2	Cups		1	5	Barrel	From 154 HP @ 7000 RPM in 1926 to 285 HP @ 8000 in 1928-29.
2	Cups		1	3	Barrel	Marine Racing Engine, 160 HP @ 4000 RPM, Supercharged. Direct Forerunner of Offen-hauser Engine.
4	Cups		1	3	Barrel	Designed by James L. Yarian.

appendix V

THE MILLER 91-CU.-IN. RACING CAR

NOTE: + = Lockhart RD Miller, 1927   RD = Rear drive   FD = Front drive

COMPREHENSIVE SPECIFICATIONS

Engine

Cylinders	Eight, in line
Bore and stroke	2.1875 by 3.0 in. (55.54 by 76.20 mm)
Displacement	90.2 cu. in. (1,478 cc)
Horsepower	285 at 8,100 RPM (Lockhart)
Supercharge pressure	30 PSI
Firing order	1-5-3-7-4-8-2-6
Compression ratio	9 to 1
Engine weight	330 lb.
Cylinder block	Two, cast iron, four cylinders each
Cylinder head	Integral
Camshafts	Dual overhead
Camshaft drive	Spur-gear train on ball bearings
Crankcase type	Barrel, aluminum
Ignition system	Robert Bosch magneto, Type FHa8
Valves per cylinder	Two
Combustion chambers	Hemispherical, fully machined
Valve diameter	In. 1.25, Ex. 1.1875 in.
Valve port diameter	1.4375 in.
Valve included angle	94°
Valve timing	5° – 38° – 35° – 8°
Valve lift	0.34375 in.
Valve spring pressure	87 lb. closed, 110 lb. open
Camshaft bearings	Aluminum, ten per shaft

Cam followers	Piston type, radiussed and keyed
Connecting rod type	Tubular, two cap bolts
Connecting rod weight	14.5 oz.
Connecting rod bearings	White metal, poured in rods
Rod bearing diameter, width	1.6875 by 1.34375 in.
Rod bearing clearance	0.0025 in.
Piston type	Solid skirt, aluminum
Piston weight, less rings	4.5 oz.
Piston rings	Two compression, one oil ring
Piston diameter	2.180 in.
Skirt-to-wall clearance	0.006 to 0.008 in.
Wrist-pin bushing	Steel
Wrist-pin diameter	0.625 in.
Wrist pins retained by	Aluminum buttons
Crankshaft type	Four-four, counterbalanced
Crankshaft weight	46 lb.
Main bearings	Five. Front main a bronze bushing, others poured in diaphragms
Main bearing diameter	1.875 in.
Bearing width, front to rear	2.0, 1.0, 1.625, 1.0, 2.0 in.
Main bearing clearance	0.0025 in.
Main bearing supports	Bronze diaphragms spigoted into and bolted to crankcase bulkheads
Lubrication system	Dry sump, duplex gear-type pump
Oil capacity	4.5 gal.
Water pump	Centrifugal
Cooling system capacity	4.5 gal.
Carburetor	Winfield 2-in. Type SR
Supercharger	Miller centrifugal
Supercharger impeller diameter	8 in.
Supercharger drive ratio	Five to one (standard)
Spark plugs	One per cyl.—Champion J10
Spark advance	Manual
Crankshaft degrees spark advance	33°
Flywheel diameter	9.5 in.
Clutch	9-in., four-face, self-locking
Transmission	Three-speed & reverse, spur gears
Transmission weight	50 lb.
Engine mounts	Three
Transmission mounts	Three
Exhaust system	Eight-branch manifold; one 4-in. or two 2-in. exhaust pipes

Chassis & Body

Weight, complete car	1,400 lb.
Frontal area, RD	8.82 sq. ft. 11.26 including rear wheels
Length	149 in. RD 146 in. FD
Wheelbase	100 in.
Tread	52 in.
Height at radiator	37 in. RD 36.5 in. FD
Height at cowl	45 in. RD 38.5 in. FD
Height at seat	20 in. RD 11 in. FD
Height at frame	18 in.
Frame width	21 in. RD 24 in. FD
Frame depth	5 in.
Body width	18 in. RD 20 in. FD
Body weight, complete	76 lb.
Frame type, material	Channel hand-formed of .125 in. mild steel
Body material	Aluminum sheet
Fuel tank material, capacity	Welded steel, internally lead-plated; 25 gal.

Suspension	RD: Half-elliptic FD: Double quarter-elliptic front, half-elliptic rear.
Axle torque control	RD: torque tube FD: front springs.
Shock absorber type	Hartford friction
Wheel spindle type	Ball bearing
Steering type	Worm & wheel, fully adjustable
Steering gear weight	15 lb.
Steering wheel turns, L to L	1.4
Brake type	Mechanical servo, cast-aluminum shoes
Brake friction area	285 sq. in. (Hepburn FD)
Hubs	Rudge-Whitworth. RD 52 mm FD 62 mm F, 52 mm R.
Axle material, type	Chrome-molybdenum. RD: solid. FD: de Dion front axle
Rear axle weight	170 lb., incl. torque tube, brakes (RD)
Front axle weight	100 lb. incl. brakes
Wheel size, type	20 in. triple-laced wire spokes, lock rings
Tire size, type	500-20 and 525-20 Firestone Speedway
Tire inflation pressure	30 to 35 PSI
Instruments	Tachometer, water temperature, oil pressure, supercharger pressure, fuel system pressure
Brake controls	Rear brakes by pedal, front brakes by hand lever

SPECIFICATIONS OF LOCKHART STUTZ BLACK HAWK

Engine

Cylinders	Sixteen, in 30° vee
Bore and stroke	2.1875 x 3.0 in.
Displacement	181.6 cu. in.
Compression ratio	6.8 to one
Supercharge boost	28 psi gauge
Horsepower	570 at 8,100 RPM
Cylinder block	Four four-cyl. Miller 91's
Crankcase	Barrel, aluminum
Crankshafts	Two, four-four type
Transmission	Three-speed: I = 2.672 to one, II = 1.355 to one, III = one to one

Chassis & Body

Weight, complete car	2,800 lb.
Frontal area	8.82 sq. ft. Incl. shrouded rear wheels, 10.85
Length	189 in.
Wheelbase	112 in.
Tread	52 in.
Height, top of headrest	46 in.
Ground clearance	4.5 in.
Fuel tank capacity	40 gal.
Oil tank capacity	6.5 gal.
Cooling system	Water plus 80-lb. ice reservoir
Suspension	18.5 to one overall ratio
Spring deflection, total	Approx. one mile
Steering	2.66 to 1. In reserve, 2.83 and 3.0 to 1
	500-20
Turning diameter	I = 99, II = 200 III = 283 MPH
Final drive ratio	0.75 in.
Tire size	Ross duplex cam and lever. Dual drag links.
Speeds in gears at 8,100 rpm	Double-cantilever one-piece springs
Lift at 283 mph	54 lb.
Resistance at 283 mph	458 lb.

appendix VI

PERFORMANCE TRENDS-INDIANAPOLIS WINNERS

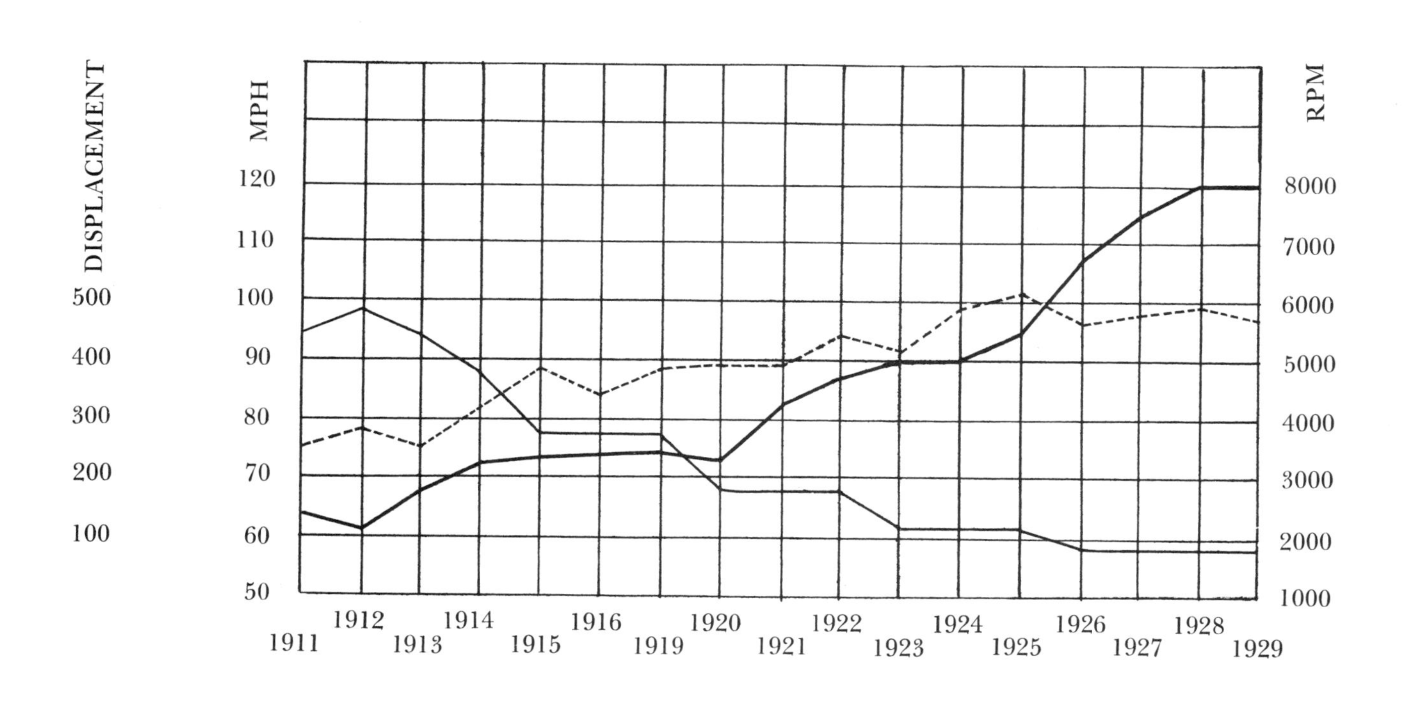

INDEX